Alzheimer Talk,

Alzheimer Talk, Text and Context

Enhancing Communication

Edited by

Boyd H. Davis
University of North Carolina – Charlotte

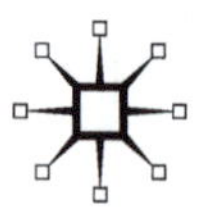

First published in hardcover 2005

First published in paperback 2008 by
PALGRAVE MACMILLAN
Houndmills, Basingstoke, Hampshire RG21 6XS and
175 Fifth Avenue, New York, N.Y. 10010
Companies and representatives throughout the world

PALGRAVE MACMILLAN is the global academic imprint of the Palgrave Macmillan division of St. Martin's Press, LLC and of Palgrave Macmillan Ltd. Macmillan® is a registered trademark in the United States, United Kingdom and other countries. Palgrave is a registered trademark in the European Union and other countries.

ISBN-13: 978–1–4039–3532–8 hardback
ISBN-10: 1–4039–3532–7 hardback
ISBN-13: 978–0–230–20694–6 paperback
ISBN-10: 0–230–20694–8 paperback

This book is printed on paper suitable for recycling and made from fully managed and sustained forest sources. Logging, pulping and manufacturing processes are expected to conform to the environmental regulations of the country of origin.

A catalogue record for this book is available from the British Library.

Library of Congress Cataloging-in-Publication Data
Alzheimer talk, text, and context : enhancing communication / edited by
Boyd H. Davis.
p. cm.
Includes bibliographical references and index.
ISBN 1–4039–3532–7 (cloth) ISBN 0–230–20694–8 (pbk)
1. Alzheimer's disease. 2. Communicative disorders. 3. Alzheimer's disease—Patients—Rehabilitation. I. Davis, Boyd H.
RC523.A3725 2005
616.831—dc22 2004065753

10 9 8 7 6 5 4 3 2
17 16 15 14 13 12 11 10 09 08

Printed and bound in Great Britain by
CPI Antony Rowe, Chippenham and Eastbourne

Contents

Acknowledgments

Friends of UNCC and especially Carol Douglass, for support on developing the host collection, Charlotte Narrative and Conversation Collection, online at http://www.newsouthvoices.uncc.edu

John Gretes, Professor of Educational Technology, University of North Carolina – Charlotte.

Ruth Greene, Chair of Psychology, Johnson C. Smith University, Charlotte, NC, for obtaining interviews from "Pleasant Homes" in the JCSU area of Charlotte, NC.

North Carolina Council on the Humanities, for support in organizing individual and community stories materials for the host collection, Charlotte Narrative and Conversation Collection.

Larry Watts, Chaplain; Sandra Adcock, Oakbridge Terrace Activities Director; and Steve Messer, Director, Plantation Estates, Charlotte, NC.

Stephen Westman, Digital Librarian; Pat Ryckman, Reference Archivist, and Robin Brabham, Curator, of Special Collections, J. Murrey Atkins Library, UNC – Charlotte: see http://www.newsouthvoices.uncc.edu

UNCC Faculty Grants Program, for subsidizing work by Boyd Davis, Linda Moore, and Dena Shenk, 1999–2003.

Chris Collins, Consultant in Old Age Psychiatry, Princess Margaret Hospital, Christchurch, New Zealand.

Matthew Croucher, Consulting Psychiatrist, Psychiatry Service for the Elderly, Princess Margaret Hospital, Christchurch, New Zealand.

Gina Tillard, Clinical Director, Department of Communication Disorders, University of Canterbury, Christchurch, New Zealand.

Notes on the Contributors

Ann P. Anas is Research Coordinator, Communication and Aging at McMaster University in Hamilton, Ontario, Canada. For the past ten years, she has coordinated S.H.A.R.E. (Seniors Helping Advance Research Excellence), a group of approximately 300 adults, 60 years and older, who participate in various communication and aging research projects at McMaster University.

Cynthia Bernstein, Professor of Linguistics, University of Memphis, works with dialect variation and narrative. She has (co)edited *The Text and Beyond: Essays in Literary Linguistics* (1994) and *Language Variety in the South Revisited* (1997).

Jeutonne P. Brewer, Associate Professor Emeritus, University of North Carolina at Greensboro, is a specialist in regional varieties of American English, with an NPR radio special on the ex-slave narratives. Her recent work emphasizes analysis from her own collection of recordings of family members with dementia.

Kerry Byrne is a doctoral candidate in Rehabilitation Sciences at University of Western Ontario, with a particular interest in empirically based communication enhancement education and training programs for spousal caregivers of individuals with dementia.

Boyd H. Davis, Cone Professor, Applied Linguistics, University of North Carolina – Charlotte, studies language change and constrained discourse: see *Dimensions of Language* (1994) and *Electronic Discourse* (1997, with J. Brewer). Articles keyed to her digital corpora of regional English, including cognitively impaired speakers, appear in *Journal of Aging Studies* and *Geriatric Nursing*.

Nancy Green, Assistant Professor of Computer Science, University of North Carolina at Greensboro, has published articles on her research on dialogue in journals such as *Computational Linguistics* and *Discourse Processes*. Her current NSF-sponsored project is on use of artificial intelligence to produce patient-tailored information in genetic counseling.

Heidi E. Hamilton, Associate Professor of Linguistics, Georgetown University, has published *Glimmers: A Journey through Alzheimer's Disease* (2003), and co-edited *The Handbook of Discourse Analysis* (2001). Her other books include: *Language and Communication in Old Age* (1999), and *Conversations with an Alzheimer's Patient: An Interactional Sociolinguistic Study* (1994).

Margaret Maclagan, Associate Professor, Communications Disorders, University of Canterbury, New Zealand, studies language change over time, focusing on the pronunciation of New Zealand English (*New Zealand English: Its Origins and Evolution*, with Gordon, Campbell, Hay, Sudbury and Trudgill, 2004) and Maori.

Peyton Mason heads Linguistic Insights, Inc. As an independent scholar in sociology, he spent twenty years in survey and marketing research, and currently applies techniques from corpus and text analysis to articles on oral and electronic focus groups, interviews, and depositions; a recent article with Davis is in *Discourse Technology* (Georgetown University Round Table, 29).

Linda Moore, Associate Professor of Nursing, University of North Carolina – Charlotte, is a Geriatric Nurse Practitioner interested in critical care in the elderly. She co-authors *Pharmacology for Nursing Care*; recent articles appear in *Geriatric Nursing* and *Critical Care Nursing Clinics of North America*.

Guenter M.J. Nold is Professor and Dean at the University of Dortmund, in English and American Studies, and is a member of the Scientific Consortium implementing DESI, the national study of performance-levels of German students. Currently he compares learning and retention in bilinguals with communicative and cognitive impairments.

J.B. Orange is Associate Professor in Communication Sciences and Disorders at the University of Western Ontario. His most recent articles on discourse, conversation, and pragmatic analyses of persons with various forms of dementia appear in *Brain and Language, Journal of Neurolinguistics, Journal of Applied Communication Research, Topics in Language Disorders*, and chapters in many books.

Charlene Pope is Assistant Professor in the Medical University of South Carolina College of Nursing and College of Health Professions. Her interests are in racial, ethnic and linguistic variations during provider–patient communication. She serves on an advisory committee for the Office of Minority Health, for whom she produced a paper on Language Access in Nursing.

Danielle N. Ripich, President of the University of New England, won the NSA Best Clinical Practice Award for a caregiver training program based on her studies (i.e. *Brain and Language, Clinical Gerontologist*) on features of language used by Alzheimer's speakers. She is developing adult language assessment tools that incorporate gender and culture.

Lisa Russell-Pinson is Project Manager of Project MORE, where she oversees the development of a corpus of oral narratives at University of North Carolina – Charlotte, and is Adjunct Assistant Professor in the TESOL Program. Her research is on medical discourse, and her dissertation is on *Variation in Medical English Writings*. Her recent work emphasizes corpus linguistics approaches.

Ellen Bouchard Ryan is Professor of Psychiatry and Gerontology at McMaster University, Hamilton, Canada. Her articles on the intersections of communication, aging and health, including studies of attitudes towards aging, have appeared in the *Journal of Language and Social Psychology, International Journal of Aging and Human Development, Psychology and Aging*, and *Journal of Gerontology*.

Dena Shenk, Professor of Anthropology, directs the Gerontology Program at the University of North Carolina – Charlotte. Her current publications are on aging, diversity and person-centered care: *Someone to Lend a Helping Hand: Women Growing Older in Rural Minnesota*; articles in *Aging and Identity, Journal of Women and Aging, Ageing and Society* and *Journal of Aging Studies*.

Hendrika Spykerman is a graduate student in the Sociology program at McMaster University, Hamilton, Canada, and is particularly interested in communication patterns between physicians and clients with dementia. She also has many years' experience as a nurse working with persons with dementia in long-term care facilities.

Introduction: Some Commonalities

The notion that a person with Alzheimer's disease (AD) can learn and can benefit from intervention is a recent one: As Bayles and Kim (2003: 329) note:

> widespread pessimism about the value of cognitive interventions discouraged many researchers from exploring their value and likely contributed to the Medicare cap on the amount of money that can be spent on the rehabilitation of adults with neurologically based communication disorders....In the last few years the value of behavioral interventions for improving the functioning of AD patients has increasingly been explored.

A common thread in the chapters that follow is that the AD speaker retains a number of communicative competencies which may or may not be identifiable in clinical studies, and that these competencies can be built upon in ways that benefit the persons with AD and their caregivers by supporting interventions to enhance communication.

The collection includes contributions from scholars in several fields: applied linguistics, gerontology, geriatric nursing, computer science, communication studies and communications disorders. All share a common concern, if not always the same terminology: a concern with how people with Alzheimer's Disease and their unimpaired conversation partners produce, interpret and understand language used in specific contexts and situations. Accordingly, the articles incorporate findings from multiple disciplines and draw on interrelated research areas such as aphasia or dementia.

A second commonality among the contributions is that, regardless of field or discipline, the various authors draw from, reference, or give context to a longitudinal digital corpus of spontaneous conversation and narrative elicited in natural settings from speakers with moderate to late AD; the authors utilize or reference other collections of speech and writing as appropriate. The corpus began in the winter of 1999, with conversations collected initially from approximately one dozen residents in the Alzheimer's Unit of "Pleasant Meadows," a private, not-for-profit retirement living community in the Charlotte, NC area that includes assisted living, hospital facilities, and an Alzheimer's wing. Conversations have been collected once or more times a month, depending on the interest and availability of the residents. In 2002–3, more conversationalists were added, including six from "Pleasant Manor," an assisted-living facility, and three attending "Pleasant Adult Day Care;" conversations with all of these and subsequent conversation partners, including those added since 2003, have continued as their health permits. The corpus has been created and is being made available on the World Wide Web, projected for 2005. Its availability will allow researchers to:

- exemplify the potential for empirical analyses of language usage;
- extend the work of researchers on Alzheimer's discourse across the disciplines by supporting the development of communicative interventions, and training in such interventions, and
- complement clinical analyses.

The corpus of conversations

For this collection, most of the contributors have drawn on a shared set of two or more years of conversations with four of the residents at Pleasant Meadows: "Aileen Copeland" (early and early moderate AD), "Glory Mason" (early moderate and moderate AD), "Robbie Walters" (moderate AD) and "Larry Williams" (late moderate to late AD). All four are white, first-language speakers of English, in their 70s and 80s; two are from the region and two are from outside the region. The four residents were identified by staff at Pleasant Meadows as being people who would enjoy conversation and visits; the attribution of AD was shared with us by residents or their relatives. A site-condition on our visits was that we not give full testing batteries consisting of multiple tests, or consult with physicians: most residents were unwilling or unable to complete the one cognitive test (Seven-Minute Screening: Solomon *et al.* 1998) that we were able to work fairly naturally into conversation, and we discontinued them almost immediately. A working assumption of our visits was that in many cases, new staff, some family members, and many visitors would know no more than we, so that our efforts to initiate, respond, and sustain conversation over time would be reasonably representative of new staff.

The additions from the larger collection are as follows: Brewer adds conversations from her own corpus, collected in the late 1980s, in a family setting, while Nold adds conversations from the larger corpus with a bilingual (German-English) woman, and Pope and Ripich add conversations with African-Americans from Pleasant Manor.

The corpus of conversation with people who have cognitive impairment, particularly Alzheimer's, is a subset of the Charlotte Narrative and Conversation Collection (CNCC), which includes approximately 700 oral narratives, interviews and conversations. This corpus represents speakers from greater Mecklenburg County, North Carolina, and adjacent counties, by embodying the varied ethnicities and languages of the region, including multiple varieties of English, Spanish, Chinese, Arabic and other languages spoken in the area. The CNCC attempts to create a regionally representative collection of the ways of speaking in a New South city in the US, at the millennium. It contains many pieces of people's lives, as the speakers draw on those parts for conversation or temporary reminiscence. The corpus is being made fully web-accessible for audio/video and transcripts. Transcripts and audio/video are encoded using the Text Encoding Initiative guidelines, tagged for metadata according to the Dublin Core principles. (For further details, consult the host site, New South Voices, at http://www.newsouthvoices.uncc.edu, the American National Corpus, and Sperburg-McQueen and Burnard 2002).

Many transcripts of the component of cognitively impaired speakers will be web–delivered through the larger corpus at the CNCC site, with aliases for names of all speakers, anonymization of details as appropriate and password-protected access for researchers to audio and, when available, video. To date, about 20 percent of the conversations have been coded for discourse features. In addition to its use for this collection, the available transcripts of narratives and conversation from this component of the CNCC are currently being used for intensive interdisciplinary courses for graduate and undergraduate students; graduate and undergraduate student capstone research projects, and workshops and seminars at regional sites.

Having a web-accessible corpus allows researchers from many disciplines, including those represented by this collection, to apply empirical methodologies for qualitative and quantitative analyses of digital corpora. Researchers can examine social interaction and collaboration with AD speakers with respect to conversation, discourse and narrative; because the audio is digital, scholars can look at issues of timing for delay in responses or topic-initiation, maintenance, or change. Researchers can draw on the larger corpus, the CNCC, for conversations and narratives by non-impaired speakers with similar ages and demographics; accordingly, they can examine both real and apparent change in language over time (Labov 1994; 2001). Researchers can incorporate current research on coherence and cohesion, both at sentence-level and beyond, to conduct empirically based investigations of retention of social and pragmatic language competencies and strategies by AD speakers. Such investigations can ground the development and testing of communicative and educational interventions.

Finally, this corpus, and the design for the present book, the first collection of chapters based on it, responds to the needs identified by Orange (2001) who calls for the development and implementation of "empirically sound caregiver communication enhancement education and training programs.... including:

(a) how best to optimize residual communication abilities of individuals with AD;
(b) how to capitalize on communication strengths and problem-solving skills of family caregivers;
(c) targeting challenging behaviors of individuals with AD that are related to language and communication problems and that may be contributing to increased levels of caregiver stress, burden, and depression; and
(d) responding to changes in language, communication, and cognition over the course of AD (2001: 243–4)

Before we can optimize abilities, we must describe them more precisely, particularly since we need to understand variation and fluctuation in language production, fluency and communicative force over time. For example, we need to be able to answer questions such as: how can we inventory conversation sequences? Identify narrative cues? Become more confident in our analyses of register variation or style/code switching? Our collection attends to such questions and to underlying, sometimes unquestioned, assumptions about the AD speaker with regard to language, conversation, and discourse and socially based identities.

While not making claims for the linguistic performance or competence of any individual speaker based on a corpus of naturalistic *usage* by a set of speakers (see Newmeyer 2003), the following can be said about the corpus: (a) that it is a reasonable sample of naturalistic and quasi-naturalistic usage by men and women with varying degrees of AD; (b) that this sample suggests what usage features may be assumed to be retained and available for conversation partners to use in support of conversation and potential communicative interventions; and (c) the sample of usage is a valuable adjunct to clinical testing which has been largely concerned with identifying preserved features of linguistic performance.

A corpus of AD discourse produced by individuals with AD and partners lets us question a number of assumptions about language and the communicative abilities of impaired speakers (see Goodwin 2003). The Alzheimer's speaker, unlike the speaker with either Broca's or Wernicke's aphasia or traumatic brain injury, has diffuse deterioration. We do not as yet have sufficient information to be able to claim that a particular speech omission suggests the location of a lesion. For example, in a two-clause sentence some speakers are unable to produce the phrase following the verb in the second clause, others omit the verb itself, others are unable to produce the noun subject. A corpus of usage can help us see if the characterizations of Alzheimer's discourse by earlier researchers need modification or expansion; as Hamilton (1994: 6) comments:

> It is only by teasing apart the variety of influences underlying communicative strengths and weaknesses in real-life interactions that we may come closer to understanding the extent to which the discourse strategies of healthy conversational partners augment or offset the seemingly relentless decline of these patients' communicative abilities.

Summary of contents

The chapters in the two sections of the book are designed to talk to each other from the different disciplinary perspectives brought by the authors; common themes are highlighted by Hamilton's Epilogue. This is neither a collection of clinical studies of language and dementia, nor is it a collection exclusively of studies of communicative interventions, although work by scholars in both areas is consistently referenced, and one goal of the collection is to contribute to the development of interventions. Instead, the chapters in this book are keyed to naturally occurring conversations over time. They augment collections such as those edited by Lubinski *et al.* (1991) or Cherney *et al.* (1998) by presenting data-rich examples which illustrate discussions of discourse production analysis directed to speech-language pathologists. The articles respond also to a more general context for language and aging, including how language in aging interacts with identity, social practices, and intergenerational issues, as suggested by multidisciplinary collections edited by Hamilton (1999) and Hummert and Nussbaum (2001). A number of the chapters continue the sociolinguistic perspectives illustrated earlier by Hamilton (1994; cf. Ray 1996) and Ramanathan (1997; cf. Meijilson 1999), complement the pragmatic and interactive perspective illustrated by Goodwin (2003), and focus on talk-in-interaction.

Talk and text

As the ability to impart information diminishes, the social-relational features of conversation become all the more important. The first section of the collection includes emphasis on identities across the life course; personhood; social identity; social relationships; caregiver concerns; and a potential strategy for communicative intervention.

Dena Shenk uses a life-course perspective, to review how choices and decisions over time affect transitions among roles, and then to exemplify how two of the Alzheimer's speakers from the corpus may be seen as presenting their different selves in the larger context of their various choices across a lifetime. Following scholars who note that the "self" is more than memory, she is particularly interested in demonstrating how, in conversations with their visiting Conversation Partners, AD speakers display identity roles and share their fears about a potential loss of self. Charlene Pope and Danielle Ripich select examples from the corpus to explore another, if related, perspective on identity: social identity as constituted and co-constructed in conversational interaction. They review speaking variations related to gender and ethnicity, presented both by persons with AD and by their unimpaired partners in conversation. Their study of the role of social variation in talk with AD speakers underscores the need for further study by health and human services workers of the various consequences of speaking practices and speaking styles.

Ellen Bouchard Ryan, Kerry Byrne, Hendrika Spykerman and J.B. Orange use the corpus to focus on particular communicative strategies, used in a series of conversations with one Alzheimer's speaker. The strategies are arrayed to demonstrate how they meet personal needs identified by Kitwood (1997); they highlight the efficacy of positive care interactions for affirming the personhood of the speaker with dementia, reinforce the retention of communicative competencies, and call attention to the active, creative contributions of the speaker with dementia.

Texts can come in two tongues: Guenter Nold's ongoing work with the didactics of second language acquisition led him to explore features from both first and second languages that Alzheimer's speakers retain and can employ. His first corpus-based study is of his conversations with a woman whose life in America has caused the replacement of her mother tongue, German, by English: Nold's conversations seek to identify what of her first language she retains. His second case study examines his efforts to retrieve German lexical items from an English-speaking man who had learned some German in his youth, through a conversation that incorporates second-language teaching activities.

Jeutonne Brewer's in-laws lived with Jeutonne and her husband, Chris. In 1988 and 1989, Brewer recorded a series of conversations with Ruth, her mother-in-law, who by then was experiencing considerable cognitive difficulties and was assumed to have AD. Brewer has written earlier analyses of Ruth's repetition. Here, she selects excerpts from Ruth's conversations to illustrate the "carousel conversation" experienced by caregivers and to highlight the personal needs the Alzheimer's speaker may signal.

Boyd Davis and Cynthia Bernstein focus on one specific speaker in the corpus to examine how the impaired speaker sustains relational goals in discourse when the ability to produce information is reduced. Their analysis of specific features in the conversational interaction is linked to two threads of research: that which shows the Alzheimer's speaker increasing their use of pragmatic constructs as their language declines, and that which suggests rapport, relationships and positioning as crucial components of conversation with impaired speakers. Later, Davis uses a review of functions of some discourse markers retained in the talk of AD speakers in the corpus, to look at the use of *so* to preface statements and declarative questions. Given that caregivers, at least in agency or institutional settings, spend a good bit of effort asking questions of AD speakers, she suggests that corpora could expand our understanding of the kinds of usage that may be easier for AD speakers to handle, and the variation they present.

Margaret Maclagan and Peyton Mason share an interest in language variation, though from different perspectives: Maclagan typically looks at New Zealand phonology, while Mason often works at the level of discourse on stance and modality. In their case study from the corpus, they look closely at different facets of one speaker's variation in lexical richness across fifty conversations, finding that analyzing spontaneous speech did not support the identification of certain kinds of patterns of deterioration: the speaker had good days for talking, and bad ones.

Text and context

A primary objective for using the corpus is looking for potential ways to enhance communications between Alzheimer's speakers and their caregivers. Kerry Byrne and J.B. Orange offer a careful review of the kinds of problems family caregivers have identified about communicating with persons having dementia, and review current communication interventions. They call for more studies with appropriate outcome measures and alternative methodologies.

Written text by Alzheimer's speakers can be important. Ellen Bouchard Ryan, Hendrika Spykerman and Ann Anas review how people with dementia use reading and writing for a variety of personal satisfactions, to argue that these activities are both possible and desirable. Their review expands empirical findings with reports of lived experience, to demonstrate the creativity and courage displayed by persons with dementia, as they use writing to maintain their sense of self. Nancy Green's work in natural language technology on computational models and analysis of dialogue, narrative, and conversational implicature grounds her discussion of the computer program she is working to develop. Elements of the program are connected to the conversational interactions of the corpus; the program will simulate conversation between a caregiver and a person with AD, so that the user – a caregiver – can practice ways to help the AD speaker converse, and to keep the conversation going.

Text about AD creates contexts: Lisa Russell-Pinson and Linda Moore are concerned with the way Internet websites establish a context for discussing AD. Although patient education is a priority, the amount of web-based information, even good information, can be overwhelming to persons and family members who think they are going to have to deal with a particular disease or condition. Russell-Pinson and Moore's review of how the flagship Alzheimer's Association site uses interrogatives to guide readers through complex information also offers implicit suggestions for writing to the lay reader.

Epilogue: the next step

Heidi Hamilton identifies four themes: the prism, the soliloquy, the couch and the dance, to discuss the evolving study of language and AD. She begins with a review of "where we've been": the state of the field in 1981 when she first began collecting conversations with an AD speaker, and ends with a call for the next generation of studies that differentiate among speakers with Alzheimer's and that expand our understanding of contexts by exploring relationships between conversational partners.

A more personal history of the project

This book probably began in 1996, with the day my mother didn't recognize me.

In the three months since my last visit, both parents had slid into dementias, unable to hide it any longer. Mother's self-proclaimed "job" had been to take care of Dad, in the way she wanted – dismissing the helpers, the bringers-of-food, the visitors and family members, hiding the weaknesses each was experiencing. Now, the neighbors had called to say he was wandering at all hours: across the golf course, through the yards, down by the lake, startling the deer. They had steadfastly refused to come-live-with; we were packing to move them, now, to the only assisted living center that would accept couples, as each had separately confided that the other had AD and would need constant companionship. When she saw the packing barrels, something broke.

"I know I know you," she said. "And I know I like you. You have such a kind face. You know me."

"Yes," I said. "Yes, I have known you for a long time and I love you. Why don't you sit there and I'll keep packing."

"I have to go but I don't want to," she said. "Maybe it won't be for very long."

Over the next two years, our family learned how crucial conversation can be for people in assisted living situations; as Dijkstra *et al.* (2004) comment, it is often the only way an older person in that situation, particularly one with dementia, can have enjoyable social moments. With my colleagues, we reviewed the research across multiple fields, on eliciting and sustaining the production and collaborative co-construction of conversation and narrative. Informally, these became my working hypotheses: (1) that both professional and family caregivers need more help in learning ways to communicate successfully with older persons as dementia increases; (2) communicative interactions need to be tucked into normal opportunities for conversation: while delivering a meal, suggesting a bath, trying to get dressed; (3) training in communicative techniques need to be delivered in small, almost transparent chunks or skills that are quick to master and easy to tuck into the daily routine; and (4) a collection of conversations over time with different people, affected by AD to different degrees, would support an empirical investigation of language patterns, with the hope that such an investigation could ground one or more possible communicative interventions. At the end of 1999, I began, with Linda Moore, to develop a corpus of conversations with Conversation Partners at Pleasant Meadows who had AD. The goal was and is to make the corpus freely available.

In 2000, Guenter Nold joined us as a regular Conversation Partner and researcher: first on his sabbatical leave, and then, as Moore was called to administrative duties, during twice-yearly visits from Germany until 2004. In 2001, Nancy Green began work on a prototype for an interactive caregiver training simulation and Dena Shenk began to examine how the conversations lent themselves to comparisons with non-impaired speakers. In 2002–3, Ellen Ryan and Margaret Maclagan began analysis of the first-draft transcripts. Ruth Greene and several of her students at Johnson C. Smith University collected and contributed a small set of normal and impaired conversations with aging African-American men and women. Maclagan began to collect parallel conversations in New Zealand English in 2003 from her home base at Christchurch, and Jeutonne Brewer added a set of family conversations recorded in Greensboro, NC, during the previous decade. In 2004, Lisa Russell-Pinson, project manager for Project MORE's augmentation of the Charlotte Narrative and Conversation Collection, began looking at discourse about persons with Alzheimer's. Speech and communications students working with Ellen Ryan, J.B. Orange and Kerry Byrne began "second edits" on a set of samples from the discourse in 2004, using the discourse coding subsystem of the CHILDES program (see further discussion and references in Ryan, Byrne, Spykerman and Orange, this volume). What began from personal experience has grown into a collection of data with national and international significance. Both the data and the analyses in this book are offered as part of our increasing understanding of the texts produced by speakers with Alzheimer's disease and contexts within which they produce these texts as they strive to retain a sense of personhood through the course of the disease.

References

Bayles, K. & Kim, E. (2003) "Improving the function of individuals with Alzheimer's disease: emergence of behavioral interventions." *Journal of Communication Disorders* 36: 327–43.

Cherney, L., Shadden, B., & Coelho, C., eds. (1998) *Analyzing Discourse in Communicatively Impaired Adults*. Gaithersburg, Maryland: Aspen Publishers.

Dijkstra, K., Bourgeois, M., Allen, R., & Burgio, L. (2004) "Conversational coherence: discourse analysis of older adults with and without dementia." *Journal of Neurolinguistics* 17: 263–83

Goodwin, C., ed. (2003) *Conversation and Brain Damage*. Oxford: Oxford University Press.

Hamilton, H. (1989) "Conversations with an Alzheimer's Patient: an Interactional Examination of Questions and Responses." PhD dissertation, Georgetown University.

Hamilton, H. (1994) *Conversations with an Alzheimer's Patient: An Interactional Sociolinguistic Study*. Cambridge: Cambridge University Press.

Hamilton, H. (1999) *Language and Communication in Old Age: Multidisciplinary Perspectives*. NY: Garland.

Hummert, M. & Nussbaum, J., eds. (2001) *Aging, Communication, and Health: Linking Research and Practice for Successful Aging*. Mahwah, NJ: Erlbaum.

Kitwood, T. (1997) *Dementia Reconsidered: The Person Comes First*. Berkshire, UK: Open University Press.

Labov, W. (1994) *Principles of Linguistic Change. Volume I: Internal Factors.* Oxford: Blackwell.

Labov, W. (2001) *Principles of Linguistic Change. Volume II: Social Factors.* Oxford: Blackwell.

Lubinski, R., Orange, J.B., Henderson, D.H., & Stecker, N., eds. (1991) *Dementia and Communication.* Philadelphia: Mosby.

Meijilson, S. (1999) [Review of Ramanathan 1997] *Journal of Pragmatics* 31: 135–8.

Newmeyer, F. (2003) "Grammar is grammar and usage is usage." *Language* 79: 682–707.

Orange, J.B. (2001) "Family Caregivers, Communication, and Alzheimer's Disease." In M.L. Hummert & J.F. Nussbaum, eds. *Aging, Communication, and Health: Linking Research and Practice for Successful Aging.* Mahwah, NJ: Erlbaum, pp. 225–48.

Ramanathan, V. (1997) *Alzheimer Discourse: Some Sociolinguistic Dimensions.* Hillsdale, NJ: Erlbaum.

Ray, R. (1996) [Review of Hamilton]. *Language* 72: 859–60.

Solomon, P., Hirschoff, A., Kelly, B., Relin, M., Brush, M., DeVeaux, R., & Pendlebury, W. (1998) "A 7 minute neurocognitive screening battery highly sensitive to Alzheimer's disease." *Archives of Neurology* 55: 349–55.

Sperberg-McQueen, C. & Burnard, L., eds. (2002) *TEI P4: Guidelines for Electronic Text Encoding and Interchange.* Text Encoding Initiative Consortium. XML Version: Oxford, Providence, Charlottesville, Bergen. At http://www.tei-c.org/Guidelines2/index.html

Part I

Talk and Text

1

There was an Old Woman: Maintenance of Identity by People with Alzheimer's Dementia

Dena Shenk

In this chapter, we examine how people with Alzheimer's disease maintain their identity and sense of self as evidenced in their conversation over time. An individual's ability to produce and retain self-identity is a requisite skill for social interaction. Researchers have demonstrated that this ability is not destroyed by the onset and progression of Alzheimer's disease itself (Kitwood 1990; 1993, 1997; Kitwood and Bredin 1992a; 1992b; Sabat and Harré 1992). The neurological impairment caused by Alzheimer's disease does, however, make it more difficult for the individual to effectively organize and sustain their various "selves." Focusing on causative factors within the social milieu, social science research is turning to studies of social interactions which may influence the progression of the disease (Golander & Raz, 1996; Kitwood, 1990, 1993, 1997; Kitwood & Bredin, 1992b; Nussbaum, 1991; Sabat & Harré, 1992; Sabat 2002) and its social causes (Kitwood 1990). Utilizing a lifecourse perspective, our focus in this chapter is on how the person with dementia retains and communicates a sense of identity by recounting memories and life experiences, personal values and views. The focus is on the lifetime of experiences and choices that the individual brings with him or her to later life and the experience of living with Alzheimer's disease.

The lifecourse perspective

Each person's journey through life can be viewed as a road map, offering many alternative routes to many alternative destinations. The pathways that develop as we age are a result of accumulated decisions and the consequences of those decisions (Atchley & Barusch 2004). This model allows us to explore how age-related roles and transitions among roles are played out over time, while keeping in mind the social context in which individuals adapt to new circumstances and social statuses (Giele & Elder 1998; Quadagno, 2005; Shenk *etal.* 2004). According to the lifecourse principle, we cannot understand a single phase of a person's life apart from its antecedents and consequences and in order to really understand a particular point in a person's life you have to take a look at the time and events before it and after it (Riley 1998). At the same time, any individual's life is intertwined or intersected with the lives of others (Riley 1998).

Our choices and the timing of our choices influence what our lives will be like as we age. For example, a woman who has children in her early twenties will likely be in her forties when her children leave her care to live independently. In comparison, a woman might experience a childless lifestyle in her twenties and thirties, perhaps dedicating herself to a career and then having children in her forties. For each of these women, not only will the balance between work and family, their relationships with their children, and use of leisure time in middle age be different, but their experiences in old age will also differ as well. Earlier decisions about childbearing influence the opportunities and responsibilities in old age when caregiving needs are likely to increase.

Utilizing a lifecourse perspective allows us to view the conversation of the person with Alzheimer's disease within the context of his or her life experiences, memories, values and views. A lifecourse perspective helps us to recognize the great diversity in the personal experience of the individual living with Alzheimer's and the importance of viewing the person with dementia within the context of a lifetime of memories, experiences and choices. It is a powerful approach as we move towards providing person-centered care for the person with dementia in order to enable the individual to retain a strong sense of self.

Person-centered care is based on the knowledge that the care environment plays a large role in determining the nature of the experience and the course of the disease. The mental powers that fail as a result of dementia are generally those related to thought and memory, whereas feelings and sociability need not be as seriously affected. With help and support, a confused person can remain in a state of well-being to a far greater extent than has otherwise been expected and maintain his/her identity and sense of self (Kitwood 1990, 1993, 1997; Sabat 2002). As Vittoria suggests, caregivers should "assume there are surviving selves in the Alzheimer's residents and endeavor to preserve, protect, support and engage those selves as the essential part of their work" (1998: 93). As

Tappen *et al.* (1999) explain: "With an understanding of personhood as more than cognitive functions measured by mental status scores, we can also search for strategies that will assist caregivers in developing ways of caring that avoid dehumanization, foster the expression of feelings, and encourage therapeutic relationships (p. 124). Through a person-centered approach, it becomes the role of others in the environment to help people with dementia preserve their sense of dignity and personhood.

Theories of self and identity

Various approaches have been suggested in the literature for understanding the conception and retention of self and identity for the person with Alzheimer's disease. It is generally accepted that an individual's identity includes both internal/personal aspects and external/public aspects. Sabat and Harré (1992) differentiate between personal selves and social selves. Post identifies the "then self" and "now self" of the person with Alzheimer's disease (1995). Small *et al.* (1998) refer to the preservation of self-identity being dependent upon internal (cognitive) and external (social) conditions and their study investigated the integrity of self (internal) and personae (external). Golander and Raz define "personal identity" as one's awareness of one's self and "social identity" as the way you are perceived by and interact with those around you. They explain that people with Alzheimer's disease certainly retain a sense of personal identity, but avenues for self-expression are commonly broken (1996). That is, the multiple personae which constitute one's social identity, and which require the cooperation of others in order to come into being, can be lost as an indirect result of Alzheimer's disease (Cohen-Mansfield *et al.* 2000). Sabat (2002) uses a Social Constructionist approach to define three categories of self:

(1) the self of personal identity,
(2) the self of mental and physical attributes, and
(3) the socially presented selves, or personae.

Chaudhury (2003) adds "imagined identities" that he describes as hoped for and possibly unmet goals. While Sabat's conception of the three aspects of identity will be utilized in the following discussion, it is useful to keep in mind Basting's conception of the personal and social selves as endpoints of a continuum (Bastings 2003).

Downs and Surr (2003) categorize the various theories of self into four categories that include cognitive, social, embodied, and biographical selves. Cognitive theories of self emphasize high-level thought – *I think, therefore I am*. Social theories view the self as a multiplicity constructed and maintained through social interaction, roles and relations. According to the embodied approach, the body is the vehicle of self in the world. For those using biographical approaches, biography and life history are central. They may be held in the unconscious or developed through the creation of the biographical self, that is, by understanding the person's life story. In the numerous autobiographical accounts by people with Alzheimer's disease that have been made available for the general public, however, the narrative is generally cleansed of the disease (Bastings, 2003). The approach that is taken towards understanding the maintenance of self by the person with dementia will be determined in part by the theoretical perspective. In this chapter I take a social view that sees the self as both protected and projected through interactions with others and displayed in conversation.

All persons, but especially persons with Alzheimer's disease, are dependent upon social interactions for the creation and maintenance of their personhood and sense of self (e.g. Atchley *et al.* 2004; Baltes & Baltes 1990; Goffman 1959). Due to the neurodegenerative effects of Alzheimer's disease, persons with Alzheimer's require special assistance in maintaining their personhood and self-identity (Golander & Raz 1996; Kitwood 1990, 1993, 1997; Sabat & Harré 1992). Preliminary research primarily in Great Britain indicates that a new paradigm of dementia care, person-centered care, is effective in rementing (i.e., reversing the psychosocial degenerative course of Alzheimer's disease), retarding, and perhaps preventing, destruction of the personhood of persons with Alzheimer's (Kitwood & Benson 1995; Kitwood & Bredin 1992b).

For each of us, maintenance of our sense of self is formulated and maintained in part through our relationships with others. As previous work has shown, people with Alzheimer's dementia are especially dependent upon others for maintaining their sense of self (Golander & Raz 1996; Kitwood 1990, 1993, 1997; Sabat & Harré 1992). For a person living with dementia, as internal "resources" are lost to the disease, communication and interaction can help retain the individual's sense of self. A key question is how the person maintains a sense of selfhood within a framework of narrowing cognitive abilities or as stated by Linda Clare (2003) how do we understand "the predicament of self in dementia"? Harris and Sterin (1999) explain that for the person with

Alzheimer's, "the self is in a state of flux, influenced by losses of significant roles, respect, autonomy, self worth, and confidence" (p. 241). Anne Bastings (2003) reminds us that the self is more than memory. "The person suffering memory loss might suffer the gradual depletion of his or her personal control over identity, but not a total loss of self" (p. 98). She explains that people with Alzheimer's disease will be as they are remembered, as people interact with them.

Conversation and the sharing of personal narratives provide opportunities for the person with dementia to construct, to convey, and to maintain her sense of self. We can assess this process in varied ways including the person's continued use of personal pronouns, and proper names, and references to past events, memories, values and personal preferences. In this chapter, we will discuss the maintenance and expression of self and identity of people with Alzheimer's dementia through two case studies from our collection of Alzheimer's discourse.

Analysis of two cases

The two cases are drawn from conversations over time with two conversation partners from the UNC Charlotte Alzheimer's database. Both of these women live at Pleasant Meadows. Aileen Copeland is a northern woman in the early stages of dementia. Our analysis is based on transcribed conversations with her over an eighteen-month period. Glory Mason is a southern woman who is quite deaf and apparently has more advanced cognitive impairment. We are not focused here on their narratives as life stories *per se*, but rather on an analysis of identity as displayed in the conversations between the person with dementia and her conversation partners who are the researchers and formal caregivers.

Aileen Copeland

Excerpts are drawn from our transcribed conversations with Aileen from January 2000 through April 2002, presented here in chronological order. The analysis is framed in the context of two themes that were apparent in her conversations: (1) how she is dealing with her cognitive losses and (2) her efforts to align her relationships with her personal values and views of herself. Note her regular and consistent use of personal pronouns in talking about herself, her use of proper names, discussions of past events, and statements of her personal preferences.

Dealing with cognitive losses and fear of loss of self

How does a person with Alzheimer's disease deal with her cognitive losses and express her fears of loss of self? Aileen struggles with these issues as she shares:

> To be honest with you, I hide it from – you know – all of the people I see. They don't know what's wrong with me, I'm sure, but –

At her pause, her conversation partner interjects: "There isn't anything wrong with you." Aileen responds: "I'm not my old self by any means."

She recognizes that she is being affected by the disease and she discusses her difficulty making close friends in her present home in comparison to her experiences in the past:

> I, I kind of understand because I've moved a lot. When my husband was alive we lived in New Orleans and Chicago and Boston and NY. So you know everywhere I went – we made friends. This is kind of an unusual situation for me; I don't know quite what to do. But I do as I said, I'm busy each night which is nice. I like the people here; it isn't that. I'm sure it's my problem.

Note that she discusses her current concerns in comparison to her previous experiences making friends through earlier moves with her husband, an example of a Self 2 characteristic (i.e. someone who makes friends easily). She remembers herself as someone who was always able to make friends in a new place and it is based on this self-perception that she is puzzled by her inability to make close friends at Pleasant Meadows. She clearly indicates her personal identity (Self 1) through her use of personal pronouns and consistent use of proper nouns for the places she lived with her husband. Aileen clearly states her personal preferences and values, regularly and consistently uses personal pronouns in talking about herself, uses proper names of geographic locations throughout the conversations, and discusses past events in relation to present events. These are markers of Aileen's Self 1 (personal identity) as she uses first-person pronouns to index her experiences, feelings, and beliefs as her own and Self 2 as she attempts to make her present experiences fall in line with her sense of herself and her values from earlier in her life. The conversations also reflect her efforts to make her social relationship with her son align with her self-identity (Self 3).

Aligning relationships with personal values

On numerous occasions during conversations with Aileen, she compares her relationship with her son and her two daughters. She is disappointed by her relationship with her son and with some of his decisions. The following conversation segment is in the context of discussing her son in comparison with her two daughters. For example, she likes her son's fiancé but retains a relationship with his ex-wife and she regularly brings up the tension between her children's lifestyle choices and her upbringing.

> Women are more sensitive to women things I think but he's a, you know he doesn't commit crimes or anything, he's a good man. But he just divorced his wife who I'm very fond of and she was a wonderful daughter-in-law. She helps me with all my tax information, you know. She's going to spend some time with my daughter when she arrives, great girl. And they're divorced and when I said to my daughter-in-law, I'm sorry about this, I think you know that and I hope it won't interfere with our – oh no no – and she's proved it; she's been a wonderful daughter-in-law.

In another conversation, she is talking about her recent visit with her daughter and granddaughter. Here, she praises her son-in-law, whose behavior matches her values (CP = conversation partner):

> CP 1: How old is that granddaughter now? Is she still in high school?
> AC: This is my youngest granddaughter and she's twelve.
> CP1: How nice!
> AC: She's a nice little girl. My daughter's husband is very kind to me, very thoughtful!

Aileen retains the basic values with which she was raised and talks regularly about having gone to parochial school and being taught by nuns. She is detailed in her recollections: her use of place names, for example, index a time as well as a space, which will ground her later life choices and signal her values. She will present these values in multiple conversations: like Antaeus, the classical Greek wrestler who regained his strength each time Hercules let him touch the earth, she is renewed when she retouches familiar, if now largely figurative, ground.

> AC: You know, I went to parochial schools for 12 years and I bet the nuns upstairs are listening to me going, "Tsk, tsk, tsk! That's not what I taught you!" . . .

CP1: Where was this?
AC: Up in Massachusetts. A little town called Medford. I lived in Medford but went to school in Malden. We walked, must have been a good mile, but we all walked together. Everyone in the neighborhood was friendly...
AC: I haven't been back there in ages, so... I don't have anyone left up there, but I'm happy here. I really like it.

In another conversation the following year, she again draws on her early life choices and values, to talk about her difficulty in accepting her son's choices, discussing her internal conflict quite cogently. The conversation partner suggests another way for her to understand and accept her son's decisions and Aileen continues:

AC: Well, I never thought of it that way, but I went to parochial school for 12 years with the nuns. And boy you just didn't sanction that arrangement at all! It wasn't happening back then as much as it is now.
CP: That's true. Why do you think it's so much more now?
AC: I think it's because women have gained more independence. You know a working mother when I was growing up was only because the husband had died and she had to work, but not just for a career. And I like this girl! But it's a conflict between what I was taught was right and wrong.

Aileen is able, then, to juxtapose her "then self" and her "now self," and to look at her own movements along her personal continuum between personal and social identities.

Glory Mason

Glory was born and raised in the Appalachian foothills of North Carolina and grew up one mile from the nearest little town. She is 92 years old, almost completely deaf, and apparently with more advanced cognitive losses than Aileen. Excerpts are drawn from our transcribed conversations with Glory from January 2000 through May 2003. Many of the conversations display her cognitive loss as she repeatedly asks whether it is time for lunch, for example, or in the following interchange where she jokes about her memory problems:

GM: I don't know what I've been talking about! How do you expect me to remember what I've been talking about?

CP: I don't.
GM: Can you do that yourself?
CP: No. No I cannot.
GM: (*Laughs*) I haven't been talking much because I don't have anybody much to talk to, but the lady over there. She doesn't want to talk *all* of the time. And I don't either!

She also utilizes explanations or statements that are "covers" for her memory loss such as her response of "nice folks" to the question about who she was eating with. She went on to explain: "There are so many people here, I can't remember all of their names." This is clearly a way of covering for the memory loss she is experiencing. The excerpts used in the following analysis, however, demonstrate her clear sense of self and maintenance of her identity when talking primarily about her family and religion.

Family

Earlier work has suggested that while all role identities deteriorate significantly for people with Alzheimer's disease, family roles retain the greatest prominence in the present (Cohen-Mansfield *et al.* 2000). For Glory, being family means caring for each other (see Shenk *et al.* 2002). She often mentions her parents and suggests the strength of their family ties with examples of the way they cared for her. When it was too muddy to walk the mile to school, her father took her and her brother in his fringe-topped surrey. Her mother's sister's family lived in "our farmhouse and we all worked together just like one big family." The day Glory left home to be married, "my daddy cooked me breakfast – he cried when I left . . ." Two months later, she added,

> Th' thing that hurt my d-daddy the most, my husband was working at the Ford plant in Michigan when I got married, an' it broke their hearts for me to move to Michigan.

Details about her husband also surface in nearly every conversation. He was a radio evangelist – she describes him in one conversation as a "Radio Christian" – who "had a building out in the backyard he called his studio where he went and prayed, 'n made his recordings." In a later conversation we learn that at the outset of his evangelism he and Glory went West, and he seems to have gone from one tiny Washington and California town to the next preaching while she raised their children.

These choices are prefigured in her descriptions of family roles and their impact on her life choices.

It took four successive conversations for Glory to be able to name all of her five children, but she talked of them often. She raised four sons and one daughter, and she recounts with pride how much they continue to care for her. Talking of her eldest son, she explains:

> When his daddy died I said, "Daddy took care of everything," an' I said, "I – I can't do it, I say, you gonna have to take Daddy's place," an' he does, he takes care of everything for me.

She expounds on the importance of her children as she continues that conversation:

> I – I thought it was hard when I was raisin' the kids when they were little but now I'm real proud of 'em I've got somebody I can trust.

She goes on to explain and to reiterate her values:

> Don't hardly know how much your kids mean to you 'til your daddy – 'til their daddy's gone.

The support and love of her family have always been important in Glory's life and she continues to express this key value as an indication of her self-identity.

Religion

Religion has also been a major part of Glory's life since she was young and is expressed as a key aspect of her identity. She testifies to her faith and talks about her first church and her Sunday School in nearly every conversation.

> [I] think about eternity and where we'll spend eternity. There's just the two places, heaven and hell . . . I'm going to heaven but the way's with the help of God . . . I've been a Christian a long time, I've been . . . and when I was a young girl I went to Sunday School all the time . . . . After Sunday School we'd all gather out in the yard in the shade of a tree and sit and visit until it was time to go home.

Her husband "preached hell hot and heaven beautiful," revealing that some of the phrases in her testimonies may be remnants of his sermons.

She speaks proudly of having gone from a good Christian home to a good Christian man.

Other expressions of Glory's identity

Glory continues to discuss her preferences and feelings, including on the occasion when she was frustrated by the researchers' efforts to administer the Mini Mental Status Exam (MMSE). She clearly expressed her lack of enjoyment and frustration with the task and responded: "No I can't do all of that stuff I'm old. (*Laughter*) . . . I'm not having fun with this." She was much more responsive when the conversation partner shifted the conversation to Glory's husband and children.

Although the length of her conversations have begun to decrease, even during our most recent transcribed conversation, Glory shows evidence of retaining her personal identity. She continues to use personal pronouns to refer to herself and although confused, she presents a clear sense of self, projected into the future. On one occasion, a female staff member asked whether Glory remembered the conversation partner who'd come to visit her earlier. She misunderstood the conversation partner's name and said she couldn't "see her good enough." Yet she carried on a cogent conversation saying: "Good to see you. You're living here now, aren't you"? The conversation partner replied: "Well, I come and go. I don't live here all the time. Anyway, I will see you next week" and Glory responded appropriately: "OK. I'll plan on seeing you."

Various aspects of her identity are reflected throughout these conversations. She expresses Self 1, her personal identity, as she uses personal pronouns to talk about herself. Self 2 is displayed in her continuing portrayal of herself as someone to whom family has always been very important and as a good Christian. Self 3 is clearly represented in her pride in her relationships with her children who have continued to support her since her husband died.

Conclusion

We have utilized conversations in a natural environment to understand the creation, maintenance and retention of the sense of self and identity for people with Alzheimer's disease. As we have seen, people with Alzheimer's disease continue to express their sense of self and retain their own identities if enabled to do so by those with whom they interact and communicate. Using a lifecourse model we see that the key themes in an individual's life remain important, with current choices and patterns being derived from and based on earlier choices and experiences.

Through their preferred topics of conversation people with Alzheimer's dementia continue to express their personal values, memories and experiences through which they retain and express their sense of their own "whole" identity. Aileen's regular and consistent use of personal pronouns in talking about herself, her use of proper names of geographic locations throughout the conversations, discussions of past events in relation to current events, and statements of her personal preferences and values are all clear indicators of a strong sense of self and identity. Aileen continues to grapple with the dissonance between her personal and cultural values derived from the way in which she was raised and choices she made in her earlier life, and aspects of her current life in a long-term care facility where she has difficulty making friends.

Even Glory, who is further along in her process of cognitive decline, continues to talk about the topics of family and religion that have been important to her throughout her lifecourse. Glory has always seen herself as a strong woman who made the choice to stay at home and raise her children while her husband traveled to do his work. Later, when the children were grown, she traveled with him, supporting him in his efforts. The choices she made throughout her earlier life frame the context of her current situation of living with dementia in a long-term care facility with the care and attention of her grown children now that her husband has died. She recounts practiced memories and stories of her life experiences with the assistance of her conversation partners and continues to use personal pronouns in talking about herself, discusses past events, and states her personal preferences and values.

Stephen Post (1995) warns us that "full self-identity, made possible by an intact memory that connects past and present, should not be overvalued lest those who are unconnected from their pasts by forgetfulness be excluded from the protective canopy of 'do not harm' " (p. 3). As Anne Bastings reminds us, "If personal identity is a site of self-reflection in which agency is enacted, social identity tends to be a site where we are acted upon by others. The 'whole' self is created by and experienced on both ends of the continuum simultaneously" (2003: 98). When Glory expresses how important her parents, aunts and cousins were when she was growing up, or how important her children are to her today, she is expressing the fact that closeness with family has been a key value throughout her lifecourse. In this way her personal identity (Self 1) and set of personal attributes (Self 2) are expressed in terms of her social identity (Self 3) and relationships with the members of her family and support system.

In attempting to understand how the person with dementia retains and exhibits her sense of identity, we must focus on the entire continuum of identity from personal through social. The fact that the person with Alzheimer's disease requires help to "quilt" her story because her memory is not intact does not mean she has lost her sense of self or her personal identity (Moore and Davis 2002). It means rather that she might require more assistance in communicating that identity. Perhaps the real issue is that we need to reconsider our primary cultural emphasis on independence and focus instead on interdependence and community in living and communicating with people with Alzheimer's disease.

The lifecourse perspective enables us to put the emphasis on the individual person – based on a lifetime of choices and experiences and identity through time – rather than on the disease. This is the message we hope to impart to formal and informal caregivers – to hold onto the person and help the individual to retain aspects of her identity rather than focus on the losses and decline caused by the disease. The lifetime of choices and experiences frames the experience of the person living with dementia. For example, consider our earlier example of how the age at which a woman begins child-bearing will determine the age of her children when she is in need of care and affect her family relationships. Both formal and informal caregivers should consider the myriad of ways in which the person with dementia is unique, based on a lifetime of choices and experiences, and find ways to help them retain their sense of identity as they provide care and support.

References

Atchley, Robert & Barusch, Amanda (2003) *Aging: Continuity and Change.* (10th edition). Belmont, CA: Wadsworth Publishing.

Baltes, Paul B. & Baltes, M.M. (1990) "Psychological Perspectives on Successful Aging: The Model of Selective Optimization with Compensation." In *Successful Aging: Perspectives from the Behavioral Sciences*, Paul B. Baltes & M.M. Baltes (eds.) NY: Cambridge University Press.

Bastings, Anne Davis (2003) "Looking back from loss: views of the self in Alzheimer's Disease". *Journal of Aging Studies* 17: 87–99.

Chaudhury, Habib (2003) "Toward a theory of (re) discovering the self in dementia: place as a pathway". Presented at the annual conference of the Gerontological Society of America, San Diego, November 22, 2003.

Clare, Linda (2003) "The predicament of self in dementia", symposium discussion paper presented at the annual conference of the Gerontological Society of America, San Diego, November 22, 2003.

Cohen-Mansfield, Jiska, Golander, Hava & Arnheim, Giyorah (2000) "Self-identity in older persons suffering from dementia: preliminary results." *Social Science and Medicine* 51: 381–94.

Downs, Murna & Surr, Claire (2003) "Theories of self and the implications for dementia care," presented at the annual conference of the Gerontological Society of America, San Diego, November 22, 2003.

Giele, Janet Z. & Elder, Jr. Glenn (1998) "Life Course Research: Development of a Field." In Janet Z. Giele & Glenn Elder Jr. (eds.) *Methods of Life Course Research: Qualitative and Quantitative Approaches*. Thousand Oaks, CA: Sage Publications, 5–27.

Goffman, Erving (1959) *The Presentation of Self in Everyday Life*. Woodstock, NY: The Overlook Press.

Golander, Hava & Raz, Aviad E. (1996) "The mask of dementia: images of 'demented residents' in a nursing ward". *Ageing and Society*, 16: 269–85.

Harris, Phyllis B. & Sterin, Gloria J. (1999) "Insider's perspective: defining and preserving the self of dementia". *Journal of Mental Health and Aging*, 5: 241–56.

Kitwood, Tom (1990) "The dialectics of dementia: with particular reference to Alzheimer's disease". *Ageing and Society*, 10: 177–96.

Kitwood, Tom (1993) "Towards a theory of dementia care: the interpersonal process". *Ageing and Society*, 13: 51–67.

Kitwood, Tom (1997) *Dementia Reconsidered: The Person Comes First*. Berkshire, UK: Open University Press.

Kitwood, Tom & Benson, Sue (eds.) (1995) *The New Culture of Dementia Care*. London: Hawker.

Kitwood, Tom & Bredin, Kathleen (1992a) "A new approach to the evaluation of dementia care". *Journal of Advances in Health and Nursing Care*, 1(5): 41–60.

Kitwood, Tom & Bredin, Kathleen (1992b) "Towards a theory of dementia care: personhood and well-being". *Ageing and Society*, 12: 269–87.

Moore, Linda & Davis, Boyd (2002) "Quilting narrative: using a repetition technique to help elderly communicators". *Geriatric Nursing* 23: 262–6.

Nussbaum, John F. (1991) "Communication, language and the institutionalised elderly". *Ageing and Society*, 11: 149–65.

Post, Stephen (1995) "Alzheimer's disease and the 'then' self". *Kennedy Institute of Ethics Journal*, 5(4): 307–32.

Quadagno, Jill (2005) *Aging and the Life Course: An Introduction to Social Gerontology* (3rd edition). NY: McGraw Hill.

Riley, Matilda White (1998) "A Life Course Approach: Autobiographical Notes". In Janet Z. Giele & Glenn Elder Jr. (eds.) *Methods of Life Course Research: Qualitative and Quantitative Approaches*. Thousand Oaks, CA: Sage, 28–51.

Sabat, Steven (2002) "Selfhood and Alzheimer's Disease." In Phyllis Harris (ed.) *The Person with Alzheimer's Disease: Pathways to Understanding the Experience*. Baltimore, MD: Johns Hopkins University Press, 88–111.

Sabat, Steven & Harré, Rom (1992) "The construction and deconstruction of self in Alzheimer's Disease." *Ageing and Society*, 12:443–61.

Shenk, Dena, Davis, Boyd, Peacock James, R. & Moore, Linda (2002) "Maintenance of self-identity in later life: case studies of two rural older women". *Journal of Aging Studies*, 16: 401–13.

Shenk, Dena, Kuwahara, Kazumi & Zablotsky, Diane (2004) "Older women's attachment to their home and possessions". *Journal of Aging Studies*, 18: 157–69.

Small, Jeff A., Geldart, Kathy, Gutman, Gloria & Scott, Mary Ann Clark (1998) "The discourse of self in dementia". *Ageing and Society*, 18: 291–316.

Tappen, R.M., Williams, C., Fishman, S. & Touhy, T. (1999) "Persistence of self in advanced Alzheimer's disease". *Image – the Journal of Nursing Scholarship*, 31(2): 121–5.

Vittoria, Anne K. (1998) "Preserving selves – identity work and dementia". *Research on Aging*, 20: 91–136.

2

Evidencing Kitwood's Personhood Strategies: Conversation as Care in Dementia

Ellen Bouchard Ryan, Kerry Byrne, Hendrika Spykerman, and J.B. Orange

The purpose of this chapter is to highlight the communication and language strategies involved in key positive care interactions identified by Kitwood (1997a) as central to affirming personhood of individuals with dementia. We focus upon the enactment of these strategies in the challenging environment of long-term care. In these facilities, residents typically are in the moderate or severe stages of dementia; staff are necessarily task-oriented; and very little knowledge is available about the residents prior to disease onset. Communication features of the positive care interactions are illustrated through transcript selections from recorded conversations in a long-term care facility with one individual in the moderate stage of dementia. As person-centered conversations lead to reciprocity, contributions on the part of the person with dementia are also shown. The real value of the examples of positive care interactions is that they reinforce the position that individuals with dementia, even those who are in the more advanced stages, retain communicative competence and are active contributors to interpersonal relationships. Moreover, the examples serve to debunk the myth that individuals with dementia in long-term care facilities are nonfunctioning, passive communicators.

Personhood and dementia

Personhood need not depend on the capabilities of the person with dementia or on our ability to overlook the person's impairments. According to Kitwood, personhood "is a standing or status that is bestowed upon one human being, by others, in the context of relationship and social being. It implies recognition, respect, and trust" (Kitwood 1997a: 8). His definition acknowledges the interdependence and interconnectedness of human beings.

In the biomedical tradition, a well-established, but false, truism is that dementia results in the "loss of self." This reductionist viewpoint proposes that pieces of the self are lost when properties that constitute the person are lost, such as cognitive abilities or functional autonomy. Obstacles imposed by the broader social and physical environments also are important determinants of perceived disability and lessened quality of life for people living with impairment (Luborsky 1994). For example, clinical practitioners and researchers tend to rely on a proxy voice to describe these losses presumably because the "subject"/"victim" is no longer able to represent the "former self" (Sabat 2001). The dominating medical model of care for dementia can create and exacerbate excess disability through a discourse exclusively based on a deficit perspective emphasizing loss, victimization and spiraling declines.

Dementia is not always simply associated with decline and loss of function. There are also positive long-term changes. As the disease progresses, the potential for growth and contribution becomes more dependent on facilitation by others. Growth occurs in areas of coping skills, compensatory actions, creativity, spirituality, and in previously hidden areas of personality (see Kitwood 1997a; Ryan, Spykerman, & Anas, this volume). Freedom of expression and a release from previous constraints and concerns may present new sources of pleasure and satisfaction for the person with dementia. In our concerns for the tragedy of dementia and the suffering endured, we lose sight of the opportunities and the real, not just imagined, potential for the human spirit to emerge in the midst of undeniably difficult circumstances (Kitwood 1997a, b; Post, 2000).

Kitwood (1997a, b) provided a holistic view of the person who lives with impairment, a "survivor" who struggles to maintain his or her personal identity, his or her personhood as he or she is confronted with diminishing abilities. He called on caregivers of persons with dementia to return to the roots of care: to care for the person, not a disease. Kitwood and Bredin (1992) called for a change of culture in dementia care, away from the old perspectives permeated with its malignant social psychology, to a new culture where person-centered care is developed, embraced and practiced. At the core of person-centered care lies the principle that an individual's life experiences, unique personality, remaining strengths, and network of relationships must be recognized and valued (see also Harris 2002; Ronch & Goldfield 2003). The concept of personhood places a major responsibility on formal care providers as facilitators for the person with dementia. While family caregivers are continually in a position to enhance personhood, the present chapter will highlight the usefulness and importance of personhood-affirming communication for formal care providers.

In Kitwood's view, the caregiver should act as facilitator for the person with dementia. Caregivers require both skills and attitude to relate to the "person" rather than the disease. The starting point for fulfilling this demanding and pivotal role, according to Kitwood and Bredin (1992), is in recognizing that the interaction is not between one party who is "damaged" and another who is whole and perfect. The person with dementia may be more vulnerable in some ways, but the caregiver also possesses weaknesses in at least some areas of functioning – perhaps on the interpersonal level, or with specific fears and uncertainties regarding impairment and death. The personhood of individuals with dementia is replenished continually through the generation and/or sustenance of interactions that are positive, stable, and secure. Such interactions meet five fundamental personal needs which overlap, coming together in the central need for love: comfort, attachment, inclusion, occupation, and identity (Kitwood 1997a).

Communication and dementia

Much research exists surrounding the language and communication difficulties experienced by individuals with dementia. Reports from caregivers about word-finding difficulties and socially inappropriate, repetitive or disruptive vocalizations (Hallberg *et al.* 1993) and empirical evidence (Kempler 1995; Ripich & Ziol 1998) reveal that lexical and pragmatic areas are largely affected. Individuals in the moderate to severe stages typical of residents in long-term care have a discourse that contains semantically empty phrases, incomplete sentences, paucity of ideas, reduced linkage between intent and wording, and lack of self-corrections (Bayles *et al.* 2000; Causino Lamar *et al.* 1994; Santo Pietro & Ostuni 2003). These deficits make it difficult for a listener, especially one who has not known the individual well, to follow the person's train of thought enough to keep up a satisfying conversation.

Consequently, conversing with an individual with dementia necessitates language and communication accommodations on the part of the interlocutor. Numerous individual language strategies have been suggested in addition to systematic attempts to train both family caregivers (see Byrne & Orange, this volume) and health care professionals (Ripich & Wykle 1996; Santo Pietro & Ostuni 2003). Language and communication strategies may change based on a variety of factors such as the severity of impairment, stage of the disease, the nature and quality of relationship, and purpose of the interaction. However, the notion of personhood as an underlying philosophy for communicating with individuals with dementia is an approach that remains unchanged across these variations.

Despite impairment, with proper support, individuals with dementia are still able to engage in meaningful conversation (Sabat 2001; Tappen *et al.* 1997). Reliance on standardized testing techniques has resulted in a focus on the impairment of abilities, despite the reality that there are many retained communication, language and cognitive abilities (Hopper *et al.* 2001; Sabat & Collins 1999; Santo Pietro & Ostuni 2003) that can be capitalized upon during the initiation and maintenance of a conversation. Santo Pietro & Ostuni (2003) outline several abilities preserved into the later stages of Alzheimer's disease including the use of procedural memory, the ability to reach memories from earlier in life, to recite, read aloud and sing, engage in social ritual and the desire for interpersonal communication. In addition, Hopper and colleagues (2001) review cognitive-linguistic abilities that remain intact at various stages of the disease. For instance, moderate-stage individuals make meaningful statements in conversation, express needs, and reminisce about past events. The ability to receive and express nonverbal cues is preserved long after linguistic skills are severely diminished (Hoffman & Platt 1990). The following section will outline some of the ways in which communication and language can be adapted with a view to enhancing personhood while considering both the impaired and retained abilities of those with dementia.

Positive care interactions and communication strategies

Kitwood (1997a) contended that "positive person work", that which is accomplished through positive interactions, must occur continually in the care environment in order for individuals with dementia to receive high quality care. Representing a person-centered approach to care, positive interactions are those that involve nurturing, healing and ultimately meet psychological need. Kitwood described positive interactions as "warm" and "rich in feeling". The following five of Kitwood's positive interactions were chosen for examination here based on their applicability to communication and language considerations for individuals with dementia. They include: recognition, negotiation, validation, collaboration, and facilitation. These types of positive interactions are discussed within the context of interactions between an individual with dementia and communicators who do not have dementia.

According to Kitwood (1997a), recognition occurs when an individual is known as a unique person by name, profiles or accomplishments. The individuality of a person is affirmed through recognition. Negotiation refers to instances when an individual is consulted about preferences, choices, and needs. Validation refers to the acceptance of reality, and acknowledging feelings of being alive, connected and real. The core features of validating are acknowledging the reality of the person's emotions and feelings, and responding to the validity of her/his feelings. Collaboration occurs when a caregiver aligns himself/herself with the person with dementia to engage in a task and to work together to achieve a common goal. Facilitation enables a person with dementia to do what he or she would otherwise be unable to do by furnishing missing parts of the intended action.

We were unable to make clear distinctions between the positive interaction categories of facilitation and collaboration in these "conversations for conversation's sake". Thus, we decided to collapse these categories and use the term facilitation, which is a major strategy used throughout the transcripts. This decision is supported by the analyses of Savundranayagam (2000), who also could not differentiate instances of facilitation and collaboration in published transcript illustrations of successful communication with individuals with dementia.

Recognition can be accomplished through the "simple act of greeting, or by careful listening over a long period of time" (Kitwood 1997a: 90). It involves both verbal and non-verbal aspects of communication (Kitwood 1997a). Recognition can involve asking an individual with dementia how he/she prefers to be addressed, and using this in subsequent communication encounters. Nonverbal behaviors such as direct eye contact and proximity considerations (e.g., getting down to her/his level if she/he is in a wheelchair) also are essential to successful recognition. To use the positive interaction feature of negotiation, it is necessary to ask questions of the individual with dementia. Researchers suggest that specific information, such as a preference, can be best ascertained using a close-ended or forced-choice format (Veall & Orange 2001). For instance, yes/no or multiple-choice types of questions provide options and are more easily understood due to retained abilities in recognition memory (Hopper 2001; Ripich *et al.* 1999). Obtaining correct information from the person with dementia can provide valuable insights about her/his preferences and needs, resulting in individualized care provision.

Validation requires the clear expression and understanding or acceptance of feelings and emotions. Savundranayagam (2000) noted that restatements, affirmations and matching comments/associations were the most frequent examples of validation used in conversations between health care professionals/researchers and individuals with dementia. Sabat (2001) reported that in his conversations with Dr. M (a client with dementia) he validated her feelings of inadequacy by restating her concerns about the difficulties associated with her dementia. This form of reflection framed within the intended emotional perspective made Dr. M feel more comfortable (i.e., interactive in a positive sense) in future communicative encounters.

The intent of facilitation [and collaboration] is to initiate, conduct and complete a task within the context of an interaction. Task initiation and completion can encompass the use of instructions (in the form of commands – direct or indirect) designed to enhance effective communication. With regard to a commonly voiced communication strategy for individuals with dementia, it is essential for caregivers to control the amount of information that their partners with dementia maintain in working memory (i.e., number of ideas or propositions), while simultaneously decreasing demands on long-term memory subsystems and processes (Bayles 2003; Rochon *et al.* 1994). Research suggests that instructions are better understood if they are short and limited to one- or two-step commands rather than as a lengthy series of instructions (Bayles 2003; Hopper 2001). This can be accomplished by breaking instructions into single steps and allowing the person with dementia to finish the first step of the task before instructing her/him to complete the next step. Camp (2003) suggests that care providers should always demonstrate first what is to be done by the individual with dementia prior to expecting a task to be executed.

Although several researchers suggest avoiding the use of open-ended questions with individuals with dementia (Ripich *et al.* 1999; Santo Pietro & Ostuni 2003; Veall & Orange 2001), there is evidence that open-ended questions can be useful in initiating a conversation. Tappen *et al.* (1997) examined conversations between nurses and individuals with advanced AD and determined that the use of broad openings (e.g., Tell me how you are feeling today) in the form of open-ended demands were successful in eliciting meaningful conversation. The authors stated that a simple opening statement often resulted in statements from the AD individual that revealed emotion and mood. The use of open-ended questions can also enable partners to initiate a desired or required task (e.g., spoken social interaction versus daily care routine). Initiating a conversation also can be improved by choosing topics from autobiographical memory because it is a better preserved subcomponent of episodic memory that contains accessible, individualized, and personally relevant topics for those with AD (Santo Pietro & Ostuni 2003).

Conversations with individuals with dementia can be filled with silent gaps and semantically empty spoken language because of the common word-finding difficulties. Thus, the topic of conversation, or the point of utterances can be challenging to ascertain (Abbott & Orange 2001). Sabat suggests the use of a facilitative speech act termed indirect repair (Sabat 1991, 2001). According to Sabat (2001), "Indirect repair refers to inquiring about the intention of the speaker, through the use of questions marked not by interrogatives but by intonation patterns, to the use of rephrasing what you think the speaker said and checking to see if you understood his or her meaning correctly" (pp. 38–9). As well, Sabat (1991) suggests that partners should not interrupt the long pauses common in spoken output of individuals with AD because their thought may return slowly if they are not distracted by interruptions. Thus, facilitation can be achieved by allowing individuals with dementia more time to reflect and access thoughts before interrupting or offering potential words or propositions.

Conversational analysis for Kitwood's positive care interactions

From the Alzheimer's component of the Charlotte Narrative and Conversation Collection (2004; see introduction to this volume) with eight residents on a nursing home special care unit, we selected the 50 audio-recorded conversations held over a period of four years with Robbie Walters (pseudonym for an 80-year-old man in the moderate stage of dementia). It should be noted that most of the conversations were recorded in the first two years, after which Robbie Walters was reluctant to converse. These five–20-minute conversations with one or more regular conversational partners associated with the Project were transcribed, segmented into utterances and analyzed for examples of the Kitwood positive care interactions. Students enrolled in undergraduate courses in communication sciences and disorders were trained to transcribe, segment and code the discourse samples using the Codes for Human Analysis of Transcripts (CHAT) which is an internationally accepted discourse coding subsystem of the CHILDES program for analyses of discourse (MacWhinney 1995). An explanation of the meaning of the symbols in the examples is provided in the Appendix.

The context for these conversations was distinctive in that experienced communicators were interacting with Robbie Walters solely to generate conversation. The purpose of each encounter was to have a conversation, unlike in many long-term care situations where the object is to complete a specific care task. In this chapter, we aim to illustrate *conversation as care*. Conversation with an individual with dementia promotes personhood when the conversational partner shows continuing interest in the person and his/her life story, preferences, emotions and needs. The interrelated nature of the positive care interactions results in their overlap within conversation. For the present purposes, we include examples that show key features of Kitwood's positive care interactions of personhood.

Recognition

Initially, the project interviewers inquired how Robbie Walters would like to be addressed in conversation. His preference for being called Robbie was followed. In the first two conversational turns of Selection 1, the conversational partner's first three utterances recognize Robbie's uniqueness through a greeting, the use of his name, and an inquiry about how he was.

Table 2.1 Selection 1: Recognition, Negotiation, Creation

BD:	Good morning Robbie.
BD:	How are you sir?
RW:	I'm fine.
BD:	Gonna take a nap or talk to me or what?
RW:	I'm [/] I'm [/] I'm ready to do any of it I guess.
BD:	Well # what would you care to do?
BD:	It's your choice.
RW:	Uh # well # I probably would put takin(g) a nap at the top of the list.
BD:	Then you take a nap and next week I'll get here earlier # before you get sleepy.
RW:	[*Chuckles*]
RW:	**Oh Well.**
BD:	Okay?
RW:	I can [# 4 seconds] uh # do any of that almost any schedule.
RW:	It don't matter.
BD:	Well the thing is # if you're ready for a good nap # I'd take it now.
BD:	[*laughs*]
RW:	You mean things are gonna get rougher?
BD:	Naw I don't think they're [/] they're rough now are they?
RW:	No xxx.
BD:	No.

Negotiation

Many interactions with a person with dementia could potentially involve negotiation. The use of negotiation allows for a feeling of being in control, of being important, and of being valued. As demonstrated in Selection 1, negotiation may be as simple as asking Robbie if he wants to take a nap. This selection illustrates that BD enables Robbie to express and choose "taking a nap" over talking to her.

Selection 2, about how to address Robbie, is another example of negotiation. In this example, the conversation partner BD negotiates with Robbie about his preference concerning naming (recognition). With some effort, BD is able to learn that he does not want to be addressed as "Mister" but rather prefers to be called "Robbie."

Table 2.2 Selection 2: Negotiation

BD:	Mister [//] what do [//] should we call [//] what should we call you?
BD:	How do you [+/.]
RW:	I'm Robbie.
BD:	Ahh.
RW:	Robbie Wilson
BD:	Well Robbie it [//] is it [//] should I call you Mister Wilson or <Robbie> [overlap>]
RW:	<no> [<overlap] not Mister.
BD:	All right.
RW:	Not used to that.
BD:	[*chuckles*]
LM:	So you like to be called Robbie.
RW:	Well yes.
LM:	Okay.
LM:	We can do that.
RW:	Mhmm.

In Selection 3, the tape recorder was started in the middle of an ongoing topic. Robbie seems to be cold and under the impression that he has no sweater. Both conversation partners BD and LM continue to negotiate with Robbie to find a satisfactory solution. In the end, Robbie admits that "it is a little warmer out there."

Table 2.3 Selection 3: Negotiation

RW:	ice have a chance to melt off of it
BD:	(*Laughed*)
LM:	Ahhh
BD:	I know what you might do.
BO:	What?
BD:	The +/.
RW:	I [/] I [/] what ever [//] every [//] everybody all other human beings do all day.
BD:	Well this is true.
BD:	I was gonna suggest a sweater.
RW:	Hmm?
BD:	Let's look in your room for a sweater.
BD:	It is chilly.
RW:	I don't have a sweater.
BD:	Shall we take a look an see if there's something else that we could borrow?
BD:	What do you think?
LM:	Or do you want to jus you know+...
LM:	It seems warmer out here than it did in that kitchen.
RW:	Well I believe it is a little warmer out here <than there> [overlap>].
LM:	<Mmhm> [<overlap].
LM:	Uh huh
LM:	You might want to sit up here then.
BD:	By the birds.
LM:	Where you were huh?
RW:	I don't care.
RW:	Jus so it's warmer xxx.
LM:	Well let's see if it it's warmer up here.
RW:	Oh okay.
BD:	That sounds like a good plan.

Validation

In selections 4 and 5, the conversation partner (i.e., GN) responds to Robbie's feelings several times, first by validating his insecurity about task completion through encouragement. Secondly, feelings of loss experienced by Robbie about past activities are explored and acknowledged.

In Selection 4, GN validates Robbie's feelings of accomplishments in the German language. Although Robbie does not readily express his emotions, he feels pleased with the compliment. Robbie's pleasure clearly shows in his wish for GN to "Have a good week" and to "Take care."

Table 2.4 Selection 4: Validation, Giving

RW: I hope I can catch on.
GN: Yeah you [//] you're [/] you're very good.
GN: That's great.
GN: You're doing a fine job.
RW: Thank you.
GN: I'm very pleased.
RW: Have a good week.
GN: Yes same to you
GN: Thanks very much.
RW: Take care.

Selection 5 shows conversation partner GN persisting in trying to draw Robbie out about how he feels about his life. Validating Robbie's loss of being able to be "out-of-doors" results in Robbie producing an uncharacteristically long sentence: "Anything that # keeps me from enjoying the out-of-doors <well>."

Table 2.5 Selection 5: Validation, Giving

GN: You're [/] you're having a good time?
RW: Well I enjoy it yes.
GN: That's fine.
RW: Yeah.
GN: Great.
RW: Yeah # well <a little> [/] a little less exciting as [/] as you get older.
GN: Yeah xx sometimes there could be a little bit more uh fun little more [/] # <more> [overlap>] activities.
RW: <mmm> [<overlap].
GN: Any activities that you miss particularly?
GN: Is it the fishing that you miss?
RW: Huh?
GN: Do you miss the fishing most?
RW: Uh # yeah I: [/] I like the out-of-doors.
GN: Yeah.
RW: Anything that # keeps me from enjoying the out-of-doors <well> [overlap>].
CG: <Yeah> [<overlap].
GN: Uh that's really what you are missing here, isn't it?
RW: mmh?
GN: That's something that you're missing here.
RW: Yes.
GN: When you sit in your chair and thinking about outdoor activities.
RW: Yeah.
GN: Okay.
RW: Thanks for checking.
GN: Yeah.

By definition, validation is the affirmation of feelings and emotions. Although the transcripts contained several examples of the other positive care personhood strategies, validation of particular emotions was non-existent. It is not possible to determine Robbie's premorbid emotionality. However, even though some of the topics discussed were potentially emotion evoking (e.g., family, growing up, etc.), Robbie did not often express emotion or feeling states in the course of the conversations.

Facilitation (and collaboration)

In Selection 6, conversation partner GN builds on Robbie's strengths and in the end GN enables Robbie to remember that he "liked all of them but I liked the raisin ones real well." This excerpt demonstrates how difficult it is for Robbie to remember what cookies he liked best and how, nevertheless, with the help of conversation partner GN Robbie is able to accomplish the task.

Table 2.6 Selection 6: Facilitation

GN:	Did you also have <a nice> [//] uh some uh uh nice uh <cookies> [//] Christmas uh cookies?
RW:	Yeah mother usually always baked the cookies.
GN:	I see.
GN:	Which one did you like best?
RW:	Uh.
GN:	Er were there any that you uh liked in particular?
RW:	Well.
GN:	Like butter cookies?
GN:	Or # butter made cookies?
GN:	Or was it uh more with uh nuts?
GN:	The ones with nuts?
GN:	Or with raisins?
GN:	Or did you like all of them?
RW:	I liked all of them but I liked the raisin ones real well.

In addition, Selection 7 illustrates how the use of facilitation enables Robbie to provide specific information about a much-enjoyed activity. For example "he had a bird dog" enables Robbie to provide the specific name "Shelley Berdette" and the fluent sentence: "Bird dogs were expensive things back then." In this excerpt, shared interest about hunting gives conversation partner BD the opportunity to provide Robbie with the missing parts of the conversation. As evidenced in this selection, shared background knowledge can be a special resource (Tappen *et al.* 1997).

Table 2.7 Selection 7: Facilitation

BD:	What else did you play besides baseball?
RW:	Hmm?
BD:	Did you play anything else besides baseball?
RW:	Oh a little bit of [//] all of it you know.
RW:	Umm played a little bit of basketball.
RW:	And uh [# *4 seconds*] we did a lot of # hunting and things like that on the hills
RW:	an uh xxx +...
BD:	That's pleasant.
RW:	mhmm.
BD:	What did you hunt rabbits squirrel?
RW:	Squirrel and a +...
RW:	Yeah <in the summers> [//] an in the summertime or in the other season it [/] it would be an [/] an +...
RW:	Ah we'd hunt # uh the other animals.
RW:	Whatever was in season.
BD:	Did you ever hunt quail?
RW:	Oh yes.
RW:	That was my favorite.
BD:	That's what my uncle hunted.
RW:	Yeah.
BD:	He had a bird dog
RW:	Yeah well # uh there's uh with us xxx available Shelley Berdette.
RW:	He uh # had a bird dog.
RW:	Bird dogs were expensive things back then.
BD:	Yeah.
BD:	Yep.
BD:	What kind did he have?
BD:	Do you remember?
RW:	Ah he had a pointer.
BD:	Oh they were the good ones. [# *7 seconds*]
BD:	Now quail [//] huntin' quail is not easy.
RW:	Well if you have a good quail dog of course you can <get your> [//] get [//] gain an advantage (on) their position an and uh +...
RW:	We hunted down through the plant areas you know back then near Viscos an on Santos an wherever good territory.

Role of the individual with dementia in promoting personhood

The personhood strategies discussed thus far have concentrated on the caregiver or conversation partner as the facilitator for the individual with dementia. However, Kitwood (1997a) suggests there are instances when the individual with dementia contributes to an interaction in a more primary fashion. In this situation, the individual with dementia takes the lead and the caregiver affirms the interaction. Kitwood (1997a) identifies *creation* and *giving* as two common examples of this sort of interaction. Creation occurs when the individual with dementia "offers something to the social setting, from his or her stock of ability and social skill" (p. 92) while giving is the act of expressing concern or affection, or making an offer of help (Kitwood 1997a).

These two concepts have not been previously explored from a purely communicative perspective, but Kitwood and Bredin (1992) identify "indicators of relative well-being," that is, abilities that can be shared by the healthy and those with dementia and include, among others, the ability to initiate social contact, humor, and show pleasure. Sabat (2001) provided empirical evidence of indicators of relative well-being by exploring the case of an individual with Alzheimer's disease (AD), Mrs. F. He demonstrated that even though she was experiencing decline in a number of cognitive areas, she still evidenced numerous indicators of relative well-being such as expressing a wide range of emotions, asserting desire or will, and the ability to be humorous. He pointed out that if one were to base interactions with Mrs. F. on the decline associated with AD then one would not recognize the abilities and attributes that have remained intact despite the disease.

Selections 8 and 9 are examples of Kitwood's concept of creation whereby Robbie spontaneously provides content to the conversation based on his humorous statements. In both of these examples, he offers "one-liners" that are extremely witty (Selection 8, "Wasn't intended for the rest of the day was it?" and Selection 9, "Oh I thought maybe I slept in the woods. I'm sure glad it wasn't that"; "You don't know who your neighbors are."). As well, in Selection 1, Robbie responds to conversation partner BD's advice to take a nap now if he is ready by stating: "You mean things are gonna get rougher?", again demonstrating his ability to offer humor to his conversational partner.

Sharing humor in conversation involves bonding and emotional closeness and, in these examples, provides an opportunity for Robbie to affirm his personhood through his ability to make people laugh.

Table 2.8 Selection 8: Creation

LM:	What do you like?
RW:	Well I like bacon.
RW:	I like about any breakfast.
RW:	I'm a breakfast man.
LM:	You're a breakfast man?
BD:	I am too.
BD:	I don't care what it is.
BD:	I'll eat it for breakfast but not for the rest of the day
LM:	Mmm.
RW:	Wasn't intended for the rest of the day was it?
LM:	[*Laughs*]
BD:	[*Laughs*]
RW:	Ahhh.

Table 2.9 Selection 9: Creation

LM:	We saw you nappin a minute ago.
RW:	Where about?
LM:	Right here.
LM:	You were asleep when we come in the room.
RW:	Oh I thought maybe I fell asleep in the woods.
RW:	I'm sure glad it wasn't that.
LM:	[*Laughs*]
BD:	I would much rather go to sleep in the woods quite frankly.
RW:	Huh?
BD:	It'd be more fun in the woods.
RW:	You don't know who your neighbors are.
BD:	Well that's true.
BD:	That is true.

In Selection 10 Robbie and his conversation partner GN are working on a few activities together involving counting and identifying pictures on cards. It begins with GN validating Robbie's attempts to answer questions about the activity and ends with a demonstration of Kitwood's notion of giving. That is, Robbie states his gratitude to GN, demonstrating a clear ability to express his appreciation. Also, in Selection 5, Robbie says "Thanks for checking" after GN validates Robbie's missing the outdoors/longing for a past activity. Both selections exemplify the ability of the individual with dementia to "give" in the conversational encounter.

Table 2.10 Selection 10: Giving

GN:	Right.
GN:	Good.
GN:	I like it.
GN:	Very good!
GN:	You remember a lot!
RW:	I appreciate it [//] your stopping by.
GN:	Yeah.
GN:	I like it [//] <I like coming> [overlap>].
RW:	<and giving> [<overlap] giving some of your time.
GN:	Yeah.

Humor and the expression of gratitude are only two possible mechanisms through which individuals with dementia can contribute meaningfully to conversations, making explicit affirmations of their own personhood in the process. Analyzing the conversational excerpts presented herein leaves one with the strong impression that Robbie, as a person living with dementia, is still capable of communicating his desires and feelings, and equally important, is able to experience some form of personal growth, even in the face of cognitive decline.

Conclusions and implications

The interaction between staff and residents with dementia in long-term care has been demonstrated to be minimal (Hallberg *et al.* 1990; Ward *et al.* 1992) with various reasons cited for limited interaction. Kitwood (1997a) recognizes the time constraints that exist for staff in long-term care facilities and suggests that interactions do not necessarily need to be of longer duration, rather each interaction needs to be of higher quality. The positive care interactions discussed in this chapter have the potential to improve interactions in long-term care facilities by the implementation of a communicative approach based on enhancing personhood. Although task completion was not the focus of the conversations analyzed, the personhood communication strategies could be used as a means of facilitating meaningful conversation during the completion of essential tasks (e.g., personal care) (Souren & Franssen 1993).

The failure to communicate with individuals with dementia in a fashion congruent with a personhood perspective may result in episodes of care characterized by Kitwood as malignant social psychology. Kitwood identifies 17 elements (e.g., withholding, ignoring, invalidation) that contribute to a malignant social psychology based on episodes of care involving individuals with dementia that he witnessed and subsequently classified. He points out that malignant does not refer to intent by caregivers, rather is a component of our "cultural inheritance". Many, if not all of these elements are affected by various components of communication and some could arguably be considered polar opposites of the positive care interactions discussed thus far. For example, imposition is identified as "forcing a person to do something, overriding desire or denying the possibility of choice on their part" (1997a: 47). Failing to use negotiation as a strategy for communication could result in imposition, as defined by Kitwood. Also, disempowerment, that is, not allowing a person to use the abilities that they do have; failing to help [an individual with dementia] complete actions that they have initiated" (p. 46), could arise if retained communication abilities are not considered and/or facilitation is not utilized during care interactions.

Discourse analyses for individuals with dementia can be used for the purposes of differential diagnosis, identifying linguistic and interactional strengths and weakness of conversational participants, monitoring disease progression, and developing or affirming theoretical frameworks of interaction, among others (Duong *et al.* 2003; Orange & Kertesz 2000). In this chapter, the discourse analyses revealed competencies of Robbie that to casual observers or conversational partners may not have been entirely obvious or viewed as a strength upon which to build collaborative, caring and rewardingly positive interactions. For example, long pauses (i.e., >5 seconds) provided by the conversational partners gave Robbie time to formulate independently his utterances and responses to questions. Moreover, the use of proper nouns rather than pronouns by the conversational partners facilitated the forward progression of "problem-free" conversation, giving Robbie the names of objects and people that helped circumvent his anomia.

While specific communication and language-based strategies for individuals with dementia have been described in the literature, one could employ personhood as an underlying philosophy in communication education and training programs for formal care providers. Indeed, we now see that the goal of affirming personhood underlies the entire positive feedback loop characterized by our Communication Enhancement Model as it applies to dyadic communication of a health provider and an older adult, with intact cognition or with impaired cognition (Orange *et al.* 1995; Ryan *et al.* 1995). In terms of educating and training, interactions that affirm personhood should be the gold standard, whether one is communicating for purposes of social interaction or to complete agenda-driven tasks. Caring for a person with dementia based on a social interactive framework that incorporates positive care personhood-affirming strategies is likely to be more rewarding than completing repetitive, task-oriented activities that only serve to shape dependence. Caregivers affirming the personhood of vulnerable individuals whose sense of self unfolds within their conversational interactions can also then participate, at least occasionally, in the warmth of reciprocity.

Appendix

Key for codes used to represent conversational features (MacWhinney, 1995)

#	= pauses of less than 2 seconds
[/]	= repetition
[//]	= retracings with corrections
<>	= marks boundaries of the targeted feature
[overlap]	= marks simultaneous talk
xx	= unclear, untranscribable word
xxx	= unclear, untranscribable words
()	= material inside parentheses are omitted by speaker but included in transcript to add clarity
"	= tag question
+...	= trailing off; incomplete utterance
+/.	= interruption; when one speaker is interrupted by another speaker

Acknowledgments

The authors express their appreciation for partial support for this research by a grant from the Social Sciences and Humanities Research Council of Canada.

The authors wish to thank Ann Anas, Boyd Davis and "Robbie Walters" for their assistance with and contributions to this chapter. We gratefully acknowledge the assistance with transcribing the conversations by Kayla Abrams, Elizabeth Bourgeois, Tracy Korzeniecki, Kim MacDonald, Coleen McAskill, and Karen White.

References

Abbott, B. & Orange, J.B. (2001, July) "Communication breakdown, topic management, and dementia of the Alzheimer's type". Paper presented at the 17th World Congress of the International Association of Gerontology. *Gerontology: International Journal of Experimental, Clinical, and Behavioural Gerontology*, 47(S1): 90.

Bayles, K.A., (2003) "Effects of working memory deficits on the communicative functioning of Alzheimer's dementia patients". *Journal of Communication Disorders*, 36(3): 209–19.

Bayles, K.A., Tomoeda, C.K., Cruz, R.F., & Mahendra, N. (2000) "Communication abilities of individuals with late-stage Alzheimer disease". *Alzheimer Disease and Associated Disorders*, 14(3): 176–81.

Camp, C. (2003, June) "Montessori and spaced retrieval: effective interventions for the cognitively impaired." 20th Annual Summer Institute on Gerontology, McMaster University, Hamilton, Canada.

Causino Lamar, M.A., Obler, L.K., Knoefel, J.E., & Martin, A.L. (1994) "Communication Patterns in End-Stage Alzheimer's Disease: Pragmatic Analyses", in Bloom, R.L., Obler, L.K., De Santi, S., & Ehrlich, J.S. (eds.). *Discourse Analysis and Application: Studies in Adult Clinical Populations* (pp. 217–35) Hillsdale, NJ: Lawrence Erlbaum Associates.

Charlotte Narrative and Conversation Collection (2004). *New South Voices home page.* Retrieved September 3, 2004 from http://newsouthvoices.uncc.edu/trial.jsp.edu/cncc.jsp.

Duong, A., Tardiff, A., & Ska, B. (2003) "Discourse about discourse: what is it and how does it progress in Alzheimer's disease?". *Brain and Cognition*, 53(2): 177–80.

Hallberg, I.R., Norberg, A., & Eriksson, S. (1990) "A comparison between the care of vocally disruptive patients and that of other residents at psychogeriatric wards". *Journal of Advanced Nursing* 15: 410–16.

Hallberg, I.R., Norberg, A., & Johnsson, K. (1993) "Verbal interaction during the lunch-meal between caregivers and vocally disruptive demented patients". *The American Journal of Alzheimer's Care and Related Disorders and Research* 8(3): 26–32.

Harris, P.B. (ed.) (2002) *The Person with Alzheimer's Disease: Pathways to Understanding the Experience.* Baltimore: Johns Hopkins University Press.

Hoffman, S.B., & Platt, C.A. (1990) *Comforting the Confused: Strategies for Managing Dementia*. NY: Springer.
Hopper, T. (2001) "Indirect interventions to facilitate communication in Alzheimer's disease". *Seminars in Speech and Language* 22(4): 305–18.
Hopper, T., Bayles, K.A., & Kim, E. (2001) "Retained neuropsychological abilities of individuals with Alzheimer's disease". *Seminars in Speech and Language* 22: 261–73.
Kempler, D. (1995) "Language Changes in Dementia of the Alzheimer Type". In R. Lubinski (ed.) *Dementia and Communication* (pp. 98–114). Philadelphia: B.C. Decker Publishing.
Kitwood, T. (1997a) *Dementia Reconsidered: The Person Comes First*. Philadelphia: Open University Press.
Kitwood, T. (1997b) "The Experience of Dementia". *Aging and Mental Health*, 1(1): 13–22.
Kitwood, T. (1998) "Towards a theory of dementia care: Ethics and interaction". *The Journal of Clinical Ethics* 9(1): 23–34.
Kitwood, T. & Bredin, K. (1992) "Towards a theory of dementia care: personhood and well-being". *Ageing and Society*, 12: 269–87.
Luborsky, M. (1994) "The cultural adversity of physical disability: erosion of full adult personhood". *Journal Of Aging Studies* 8: 239–53.
McWhinney, B. (1995) *The CHILDES Project: Tools for Analyzing Talk*. 2nd edition. Hillsdale, NJ: Lawrence Erlbaum Associates.
Orange, J.B. & Kertesz, A. (2000) "Discourse analyses and dementia." *Brain and Language*, 71: 172–4.
Orange, J.B., Ryan, E.B., Meredith, S.D., & MacLean, M.J. (1995) "Application of the communication enhancement model for longterm care residents with Alzheimer's disease." *Topics in Language Disorders*, 15(2): 20–35.
Post, S.G. (2000) *The Moral Challenge of Alzheimer Disease: Ethical Issues from Diagnosis to Dying*. Baltimore: Johns Hopkins University Press.
Ripich, D.N. & Wykle, M.L. (1996) *Alzheimer's Disease Communication Guide: The FOCUSED Program for Caregivers*. San Antonio, TX: The Psychological Corporation.
Ripich, D.N. & Ziol, E. (1998) "Dementia: A Review for the Speech-Language Pathologist", in A.F. Johnson & B.H. Jacobson (eds.) *Medical Speech-Language Pathology: A Practitioner's Guide* (pp. 467–94). NY: Thieme.
Ripich, D.N., Ziol, E., Fritsch, T., & Durand, E.J. (1999) "Training Alzheimer's disease caregivers for successful communication". *Clinical Gerontologist*, 21(1): 37–56.
Rochon, E., Waters, G.S., & Caplan, D. (1994) "Sentence comprehension in patients with Alzheimer's disease". *Brain and Language*, 46: 329–49.
Ronch, J.L. & Goldfield, J. (eds.) (2003) *Mental Wellness in Aging: Strengths-Based Approaches*. Baltimore: Health Professions Press.
Ryan, E.B., Meredith, S.D., MacLean, M.J., & Orange, J.B. (1995) "Changing the way we talk with elders: promoting health using the communication enhancement model." *International Journal of Aging and Human Development*, 41: 89–107.
Sabat, S.R. (1991) "Turn-taking, turn-giving, and Alzheimer's disease: a case study of conversation". *The Georgetown Journal of Languages and Linguistics* 2: 284–96.
Sabat, S.R. (2001) *The Experience of Alzheimer's Disease: Life Through a Tangled Veil*. Massachusetts: Blackwell.

Sabat, S.R. & Collins, M. (1999) "Intact, social, cognitive ability, and selfhood: a case study of Alzheimer's disease". *American Journal of Alzheimer's Disease*, January/February.

Santo Pietro, M.J. & Ostuni, E. (2003) *Successful Communication with Persons with Alzheimer's Disease – An In-Service Manual*. 2nd Edition. St. Louis, Mo.: Butterworth-Heinemann.

Savundranayagam, M.Y. (2000) "Personhood and effective communication with persons with dementia." Paper presented at the annual meeting of the Gerontological Society, Washington, D.C.

Souren, L. & Franssen, E. (1993) *Broken Connections: Alzheimer's Disease, Part II – Practical Guidelines for Caring for the Alzheimer Patient.* Berwyn, PA: Swets and Zeitlinger.

Tappen, R.M., Williams-Burgess, C., Edelstein, J., Touhy, T., & Fishman, S. (1997) "Communicating with individuals with Alzheimer's disease: Examination of recommended strategies". *Archives of Psychiatric Nursing* 21: 249–56.

Veall, J. & Orange, J.B. (2001, July) "Question comprehension by individuals with dementia of the Alzheimer's type." Paper presented at the 17th World Congress of the International Association of Gerontology. *Gerontology: International Journal of Experimental, Clinical, and Behavioural Gerontology*, 47(S1): 91.

Ward, T., Murphy, B., Procter, A., & Weinman, J. (1992) "An observational study of two long-stay psychogeriatric wards". *International Journal of Geriatric Psychiatry*, 7: 211–17.

3

Speak to Me, Listen to Me: Ethnic and Gender Variations in Talk and Potential Consequences in Interactions for People with Alzheimer's Disease

Charlene Pope and Danielle N. Ripich

Talk is a non-material communication tool that people use to meet their goals in social interaction. This view, held by linguistic anthropology (Duranti 1997), states that talk is a tool speakers use in various ways in differing circumstances. The successful use of talk as a tool in conversation depends on ways of speaking or practices related to one's cultural assumptions and social positions. These are learned over a lifetime and integrated by speakers as social identities (Ellis 1999). Although social practices affect communication, the role of social variations in talk with people with Alzheimer's disease (AD) has received relatively less attention than cognitive or expressive language issues in AD. Social identities are often strongly tied to expressive speaking behaviors (Reichman 1997), particularly those related to gender (Roter, Lipkin, & Korsgaard 1991) and ethnicity (Stubbe 1998). Hanks (1996) has proposed a co-constructed model of communication based on the simultaneous roles of speakers/ listeners in conversations. Within this concept of co-construction, people, as speakers and listeners, alter their communication relative to gender and ethnicity in the midst of a conversation quickly and often without conscious awareness (Jacoby & Ochs 1995). While engaged in more functional tasks, speech partners may or may not recognize references. They may index inferences into a phrase, pause, raise in pitch or other element of performance (Hanks 1996). The recognition of a common frame of social or cultural experience involves shared social identities considered an indexical ground (Hanks 1996). In all these ways, talk plays a critical role in the construction of social identity (Zimmerman 1998; Schegloff 1991; Linell 1998).

The focus of this chapter will be to discuss the basis of social identity in talk and to explore how these identities may impact interactions with persons with AD. This descriptive, sociolinguistic approach will consider ethnicity and gender primarily, but will also include discussion of social status, age factors, and institutional life as enacted factors of social identity.

Differing paradigms identify social identity as the awareness, representation, and attributions about oneself and others as unique individuals from an internally defined personal perspective, an externally acquired social self within groups, and as interactively constructed between one's sense of single and collective consciousness (Foddy & Kashima 2002), through the speaking/listening behaviors of talk. From an infant's first words in interaction (Papousek 1995) to their later encounters at the latter end of the life cycle (Saunders 1999), each person uses particular speaking practices as social actions to attach, engage, attune to, exchange information, and influence or persuade others (Holtgraves 2002). For example, we use and hear prompts everyday in participation either as paralinguistic cue to encourage participation ("*tch*" or "*um-hmm*") as well as short linguistic cue strategies such as affirmations ("*Oh yeah*"), self-disclosures ("*That happens to me . . .*"), the recognition of topic cues in turn-taking ("*you mentioned . . .*"), interruptions that show our interest ("*you go, girl*") as opposed to those that push us apart (heard in a change of subject, a correction, or indifference.. "*What-ever: : . .*"), and the endless variety of changes in pitch, rhythm or tone that make the meanings of words so much more than they are ("*You did: :* WHAT!!).

Each speech activity establishes, accepts, maintains or resists group memberships while categorizing others (Housley & Fizgerald 2002). In the course of conversation, people use often taken for granted discourse strategies in talk that accomplish multiple purposes in communication (Terrell & Ripich 1989), beyond more familiar dimensions of information exchange, relationship building, or the functional agendas that bring speakers together in caretaking or health encounters (Street 2003). Though communication accommodation theory describes how such categorizing works in relation to age and stereotyping (Coupland, Coupland, & Giles 1991), ethnicity and gender also require attention when patients with AD and their unaffected speech partners interact.

From a microsociology perspective, Derber (2000) describes how people use talk not only to affirm, maintain, and resist identities but also to negotiate for the commodity of attention according to differences in social status, since social identities are not equal in societal value. The distribution of attention may differ by social status markers, as seen in observations of ethnicity (Bailey 2000) and gender (Hopper & LeBaron 1998). These strategies or mechanisms associated with social identity in talk have been less described in comparisons among elderly populations and specifically those with Alzheimer's disease. Social identities can affect the flow of talk in a number of ways. We can examine talk as a way to learn about how individual participant speaking behaviors shift with social identities in social interactions and the consequences, especially a dimension of interest for people with Alzheimer's disease where long-held speaking practices change. Viewed over the lifetime of an individual or occurring between those of differing generations (Williams & Nussbaum 2001), social identities prompt differing speaking practices (Labov 2001) and create or reference differing contexts (Paoletti 1998).

Speaking variations related to gender and ethnicity precede memory, beginning with one's earliest socialization and habituated behavior. A toddler learns appropriate ways of speaking from how childhood caregivers and community members perceive the speaker, as seen in consideration of gender (Miller, Potts, Fung, Hoogstra, & Mintz 1990). Though comparisons of AD and non-AD speakers in interactions do not usually contrast same community member and differing member pairs, other sociolinguistic observers note that ethnic variations in speaking practices provide additional opportunities to consider how these differences act as resources in conversations (Day 1998), observed in narratives (Smitherman 1977), grammar and forms of indirectness (Morgan 1996), regional dialects and vernaculars (Rickford 1999; Wolfram, Adger, & Christian 1999) as well as in the production of potential disparities in care (Cooper & Roter 2003).

Social identities and the status differences they incur affect the distribution of attention as speakers and listeners talk (Ervin-Tripp 2001). These differences in attention can occur through documented shifts in dominance seen in topic selection or the type of questions asked (Thimm, Rademacher, & Kruse 1995; Itakura 2001), interruptions (Goldberg 1990), evaluation of particular speaking practices (Koch, Gross, & Kolts 2001), and patterns of decision-making (van den Brink-Muisnen, van Dulmen, Messerli-Rohrbach, & Bensing 2002). Differences in perceived status or power may alter the communication strategies participants use in conversation. When the sociologist Weber (1947) describes the differing forms of power, each type, whether conferred and traditional (*"Yes, my father was a Kennedy..."*), achieved and rational (deduced in the crisp consonants and long words of the over-educated or tone of authority figures), acquired by threat or coercion (*"Stop! It's the police!*), or charismatic and personal (*"I really:: would like to see:: you again*...), all use talk to influence and persuade and are sensitive to social status cues.

Communication analysis of persons with AD can identify changes in memory, word retrieval, language comprehension, processing, and cohesion based on the analysis of talk over time (Emery 1999). As a complement, more close examination with discourse and conversation analysis may identify taken-for-granted patterns of ethnic or gender-specific habituated speaking behaviors for those with Alzheimer's and their speech partners within interactions. However, traditional analytic approaches may fail to recognize social positioning in talk or its effects on the consequences of an encounter, since the markers for these more social elements are not always discernible when investigating elements such as content, cohesion or specific speech acts of the AD speaker (Ripich, Carpenter, & Ziol 2000). Specifically, the discourse strategies that subtly categorize the speech partners' social memberships (Housely & Fitzgerald 2002) may result in poorer process and outcomes for people with AD, as suggested by nonverbal communication with the elderly (Ambady, Koo, Rosenthal, & Winograd 2002). In contrast, though discourse strategies associated with social identities may continue well into early and middle stage AD, both the person with AD and their caretakers or interviewers may be unaware of the social positioning they use in talk. Since conversation is a co-constructed process between two participants with same or differing perceived social identities (Ochs 1992), close attention to social practices in talk can reveal patterns associated with gender and ethnic identity that persist and influence both people beyond their notice. These speaking and listening practices can maintain even after other dimensions of communication with AD have changed.

The next section of this chapter will look at observed discourse strategies related to social identity, particularly gender and ethnicity, and how these variations affect social interactions in persons with early stage AD. Using selections from data presented elsewhere in this volume, we will compare segments of several interviews for evidence related to shifts in social practice associated with gender and ethnicity. Talk between the person with Alzheimer's and their caretaker or interviewer were examined in same and differing gender pairs in monoracial White with White and Black with Black encounters.[1] Interviews reviewed include:

---

[1] For the purposes of this discussion, Black and White refer to members of skin-color based social groups as part of the historical social system of race in the United States for those of African-American and European-American ethnic descent that establishes a hierarchy of social status and cultural habitus (Omi & Winant 1996).

- Four White persons with AD (two female and two males) and their two adult professional White interviewers (one White female primary English speaker and one White male secondary English speaker whose primary language was German, both middle-aged). Fifteen differing sessions were reviewed.
- Two Black persons with AD (one male and one female) and their young Black interviewer (female about to enter graduate school). Eight different sessions were reviewed, with numbers based on data availability, not selection. All except the male of German origin were Southerners by dialect.

Since nursing homes are among the most segregated institutions in the United States (Howard, Sloane, Zimmerman, Eckert, Walsh, Buie, Taylor, & Koch 2002), either institutional segregation, differential access, preference or individual gatekeeping limited the availability of interracial encounters for analysis. The way language is used and the pragmatics of social variations contribute to disparities in care (Cooper & Roter 2003). Therefore, the analysis of gender and ethnic variations provides a basis for the development of communication skills training that may improve care for those with AD. Varieties of speech practices below identify situations where type, quantity, and quality of attention shifts in differing situations, with the formation of differing identity pairs. The examples address differences in address, narrative formation, topic control, and allusions to race seen in conjunction with specific speaking practices.

Openings as examples of gender and professional identities

Most of the White with White speaker encounters occur as the interviewers stroll through the public lounge or from room to room soliciting people to interview. The structure of these encounters differs from the available Black with Black speaker encounters, all of which begin in process apparently in an individual space where there are few interruptions or incursions of the others in the nursing home. Whether Black or White, the interviewers state as their purpose the collection of stories and memories for recording. Chosen as typical, the following encounter chronicles a stroll as two White interviewers speak with White patients.

IWF (Interviewer White Female)
IWM (Interviewer White Male)
with Primary White Male First (PWM1), Primary Female First speaker (PWF1), etc.
In this interview, all participants are White.

Segment 1

1 IWF: Good mor::rr:: ning.
2 Are you O-kay::: How nice to see you.
3 Good morning Mr. L (*name masked*)..... I'll check you after break-fast
4 Gonna walk around this way:: Hey::......
[*Heard changing locations, moving from one to another*]
5 Hell-oh:: Hell-oh:::: Hello Ms. B::
6 Hey... I haven't seen you in a while
7 PWM1: Huh?\
8 IWF: ..but I'm glad to see you... [*Moves on, by register addresses someone else, pitch rise suggests female*]
9 I wasn't <u>here</u> last week
10 PWF1: And I really <u>miss</u>ed you... and you're going to play
[*Next speaker is White, male, German accent second language, fluent English*]
11 IM: [*Walks around greeting.*] And so am I:::::
12 PWF2: <u>You'</u>ve been out doors:::
13 IWM: Yeah that's right:: [*Shared laughter*]
14 You been out doors (.2) too:::
15 PWF2: huh?
16 IWM: Have you been out doors too
17 PWF2: One time
18 IWM: Ah yeah.. that's great
19 Where did you go?
20 PWF2: to find out.... how cold it was
21 IWM: ah yeah... where did I... where you go:::
22 PWF2: huh/
23 IM: where did you go:::
24 PWF2: huh/
25 IWM: <u>Where</u>:: (.4) did you (.2) go:::::
26 PWF2: Just had to hear it twice (laughter)
27 IWM: I see.... I see
28 IWF: They don't ever go out..
29 IWM: I see::: Yeah she told me she went.... [*Laughter, voice changes tone as syllables become less elder-speak, addressing fellow professional, lower tone*]
[*Both move on*]

The previous example characterizes the beginning of most interactions with White patients in these samples. All were gathered in either public spaces or individual rooms in a nursing home. From line 1 to line 4, each comment greets someone in public space, yet often neither interviewer attaches nor attunes to a particular speech partner for more than a few turns. In these segments, the White interviewers introduce topics and organize the direction the topic continues in turn-taking. In line 5, the female interviewer changes her style of speaking by pitch to greet a female patient putting stress on a syllable, mirrored in cadence and pitch by her female partner as an act of attunement. Claiming a place on the conversational floor, the male interviewer moves from a peripheral listener in line 11 to share the intention of his female colleague with the indirect statement "And so am I..." referring indirectly to his intention to also be playful as the female patient says on line 10, a shared indexical ground. The second female patient initiates a topic in line 12, commenting that the male interviewer looks as if he has been outside. Offered a return statement, the female patient either does not hear or understand on line 15, a "Huh?" requesting clarification. Rather than clarify if she had difficulty hearing or did not understand the query, the male interviewer's statement becomes a question, which elicits information that the female patient has been outside one time. The use of an essentially empty expressive evaluation "Great" holds the floor as the Interviewer shifts the topic back to "where" the woman had gone. In line 20, as the female patient moves to explain "why" she went outside in an effort to maintain the topic, the male interviewer disregards the turn to pursue "where" the female patient had gone in lines 21, 23, and 25, each line becoming more exaggerated and louder in the tone considered by some to be elder-speak (Bethea & Balazas 1997). In line 26, the female patient makes a joke at their misunderstanding or at the failed turns. The male interviewer's return that "I see" does not identify anything either seen or understood. With the re-entry of the female interviewer on line 28, both interviewers alter their register to one of professional colleagues, speaking in front of the female patient, with the third person "They don't ever go out," dismissing a topic and moving away.

Admission to an institutionalized home and public availability to those in authority impacts the capacity of the elderly to speak and listen with one another. The gathering of the elderly with or without AD into a nursing home population as patients constitutes a society differing from the past experience of most people and not necessarily of their choosing. As the institution structures activities of daily life, it colonizes human experience and regulates access into lives in ways that would not have been as possible when these people lived independently. Long-term care facilities for the elderly relate to Foucault's description of the phenomenon of panopticism (Foucault 1977, 1991) as institutions monitor and control human activity. The interviewers have the power to introduce themselves into the lives of the AD patients, linger or not linger, pursue or not pursue a conversational turn, and speak as if those physically present are not there. As briefly suggested by gender mirroring in lines 9 and 10, the multiple nature of social identities offer alternatives to participation, but professional identity and the instrumental task of eliciting particular types of conversation narrow the possibilities. The power to claim the conversational floor and change topics has been associated with professional roles in health and human services as well as with gender (Ainsworth-Vaughn 1998). Each combination of gender pairs may offer differing resources to talk, but institutional roles may mediate interaction.

In these recorded conversations, it appears that gender-related roles in past life frequently direct topics for discussion. In the next two segments, a White female interviewer speaks with first a male patient (about the military and his education) and then a female patient (about the pretty Mother's Day card). The segments demonstrate other practices that occur typically in these gender pairs that are associated with topic autonomy and ways of maintaining the floor.

Segment 2

1 IWF: I'll. . . . Say.. another Good Morning ..cause you look so cheerful –
2 heh-heh
[*Chuckle, pause*]
3 I'd <u>LOVE</u>::: to talk to you about. . . . were you a young man . . .
4 in .. the <u>mil-</u> itary
5 PWM2: Young man in ..what:::?
6 IWF: In the military?\. . . Didn't <u>you</u> go to the ARM- y:::
7 PWM2: Yes::s::: [*Tone grows more certain, firm, and louder*]
8 IWF: I .<u>thought</u> . .you ..did:::
9 Did you go to the Army ..too?
10 PWM2: Well.. I::. . . . Less::. . . . say.. I . . . [*Voice is slurred*]
11 At Purdue university where I went . . .
12 they had AY div-i-shion . . . of the Ar-my::. . .
13 IWF: They had a division of the Army . . .? At that university? . . . [*Sounds as if she did not understand the slurred words well and is clarifying*]
14 PWM2: Yes.
15 IWF: What Univ-ersity was <u>tha:::at</u>?::
16 PWM2: Pur-due::
17 IWF: OH . . . PUR-Due::: That's a <u>Great Schoo-ool</u>
18 PWM2: I think it <u>IS</u>::

19 IWF: Then you were an en-gin-eer::. . . weren't you?
20 PWM2: Yes::: yes
21 IWF: Most of the Boiler makers ..(.4)
22 PWM2: Well Pur-due taught other things:::
[*Takes up the topic of education and continues*]

In terms of topic control and maintaining the conversational floor, fewer interruptions typically occur in the female interviewer–male patient conversations in these samples. This can be viewed as either power-oriented or rapport-building features (Goldberg 1990). The female interviewer is more likely to pick up topics but the male patient shifts direction in related conversational turns. Requests for information tend to be direct. Turns appear distinct with questions offered in direct formats that require specific information and yes or no answers, with time given for responses. On line 7, the male patient takes up the cue from line 6 asking about army experience, with a repair that introduces a theme for a story. Rather than prompt for a continuing story that would build a narrative, the indirect question on line 13 and the direct question on line 15 put the priority on information-gathering rather than story-telling, diminishing the opportunity for a collaborative conversational floor. With the identification of a particular school in line 16, the female interviewer in turn evaluates, with an exaggerated tone and demonstrates knowing particular information in line 21 that the male speech partner accepts and corrects in line 22. Later the segment ends when the interviewer does not pick up the identity cue about college and voices an intent to come back and talk about the military in the future and they go off to do exercise, a change of topic.

In the next segment, a White female Interviewer comments and admires greeting cards in the room of a White female patient's room as she strolls from one room to another down a hallway.

Segment 3
1 IWF: Ms D [*Name omitted*]'s going for her shower . . . [*Enters room*]
2 I LOVE this::: [=
3 PWF1: =] Idn't that . . . pretty[=
4 IWF: It's so:: pretty
5 PWF1: I wanta take that back . . . cause I feel it-that's Susan's::
6 IWF: That's who signed it..isn't it (1.0) . . . Yep (1.0) . . .
7 that's who signed it . . . (2.0)
8 "with lots of love..Susan" [*Simulated speech, reading text on object*] (1.0)
9 "Happy Mother's Day:::" (3.0) That's a beau-tiful car::d\

10 PWF1: [*Cough-clearing throat*] Is that last year?
11 IWF: I think it might be\ [*In agreement*] (1.4)
12 Well it's PAST mother's day...
13 It could have been earlier this year:: (.4) How ya Do-in'[=
14 PWF1: =]it's Mother's Da-
[*Breaks off to respond with the date, but switches to newly introduced topic in overlap*]
15 –Not so go::ood..um- as you say::..
16 I don't ..-I don't even know when Mother's Day was:: (.2)
17 heh-heh [*Chuckle*]
18 IWF: Well... stand there a minute...
19 and I'll try to figure out when it was..
20 Awright.. (.4) This is:::[=
21 PWF1: =] It's in (.2) May:: (.4) idn't it?
22 IWF: It's in May... and this is ..Sep-tem- ber... so it's been ..
23 May to June..to July to Aug-[=
24 PWF1: =] I been by my-self::[=
25 IWF: =]gust..[=
26 PWF1: [*Continues over Boyd's voice*]=] now:: ..so I don't... [*Voice trails off*]::
27 IWF: =*I* don't keep track\
28 PWF1: [*Picks up the thread*] =[..know when things are
29 IWF: Without a cal- en-dar ../I don't keep track\
30 PWF1: um-hm... Yeah:: [*Chuckle*] [*Unclear phrase*][=
31 IWF: =] But you're (.2) LOOK-in' fine :::

In the prior segment, the speakers use the resource of the collaborative conversational floor, described in other female–female encounters (Coates 1997). In response to line 2's theme of the greeting card and its "pretty"-ness, the female patient uses a rapport-building interruption in line 3 to express agreement. In the next few lines, the female interviewer uses the theme of the unknown Susan to continue focus on the card. The female interviewer seems to be using the rhythm of pauses as prompts for encouragement in line 6, when the rhythm of the seconds are counted. In the decision to not pursue the cues of parenting, children, Susan, or relationships, the interviewer returns to the topic she had opened with on line 2 of the greeting card as an object in itself. In line 10, the female patient uses a throat clearing to shift the topic to the date when it had come, beginning turns interrogating when the card had come. In line 13, when the female interviewer begins to change the subject with an inquiry into how the patient feels or is doing, the female patient tries to continue the thread of when Mother's Day occurred. Often AD patients struggle to keep up with a person with faster processing, retreating back to pick up a thread not allowed sufficient time or whose reference marks a direction not their original intention. In line 14, the patient repairs to answer the Mother's Day theme evoked in line 9, but followed quickly by the interviewer's evaluation about the card's appearance. The female patient's response interrupted self-repair in line 15 that tries to answer the well-being inquiry in line 13 continues the subject of Mother's Day in line 16. The building of a collaborative conversation floor in lines 21–29 mediates the perception of a power-oriented change of subject in line 12 with mutual responses to the Interviewer's intention to clarify date recall and the common experience of not remembering.

From lines 18 to 23, they each build collaboratively to continue the thread of trying to remember, differing in the subjects they are trying to maintain yet building on the other's thread. As a result, the female interviewer is listing the months in trying to place the date in time, while the female patient is referring to her state of aloneness as a complication in remembering. In line 27, the female interviewer finishes the intended sentence of the female patient, though not part of her conversational thread in line 20. The segment ends with a once-again gendered identification of appearance as a topic, a frequent topic in female–female encounters less observed as often in the male–male interactions. Comments on appearances as openers do not seem taken up as readily in male patient–female interviewer encounters. as seen by the opening line 1 in Segment 2 when the cue about appearing cheerful is abandoned to pursue a more gender-appropriate role of military participation for men.

Choices in topics, prosody, and collaborative floor patterns associated with gender (Tannen 1993) appear as speaking resources in the reviewed interviews, but most of the White monoracial encounters are marked more by direct and yes/no questions, evaluative meta-comments (*Great*) followed by changes of topic, and requests for information mediated more by the professional agendas of interviewers and instrumental tasks (care reviews, test-taking, etc.). This interrogative style produces fewer temporal connectives, such as *when, and then, or after,* that characterizes more narrative memories and story-telling (Ervin-Tripp & Kuntay 1997). With or without AD, institutional agendas produce institutional speaking practices that limit the potential for social interaction or response to cues based in common gender or ethnicity.

Race/ethnicity and the role of age socialization

The Black female interviewer differs from the White interviewers in age and professional rank. References throughout the interviews suggest she is in her 20s and about to enter graduate school. Based on observations, the young Black interviewer appears less accustomed to professional role autonomy and exhibits differing speaking practices compared with the older White male and female interviewers. In addition, she demonstrates gender differences in her practices with the Black male and female patients she interviews. In the following Segment 4, the female Black interviewer (IBF) speaks with a female Black patient (FBP) about where she has spent her early life. The construction of place includes cues that mark social locations for both speakers in their common memory of a particular city, Baltimore, which holds particular class connotations and significance for Blacks from one neighborhood to another (McDougall 1993), even those of differing ages.

Segment 4

1 IBF: Are you <u>from</u> (.) Char-lotte::?
2 FBP: Yeah ..this is my home.... I was born and raised here
3 IBF: Born and raised?
4 FBP: ..And I went to (.2) Balt-i-More:::
5 And I lived there for 30 years or so 'fore I come back <u>home</u>
6 IBF: [*Louder and more animated*] Did you <u>like</u> Baltimore? I was [=
FBP: =] [*Interrupts to pick up Baltimore thread, not waiting for interviewer's experience*]
7 I liked it fine\
8 IBF: =] I- I just spent the summer there[=
9 FBP: =[I was <u>raised</u>-da:: (1.0)
10 Some of my children were small (.2) when I went
11 IBF: um-hm [*With the paralinguistic cue, she puts aside her experience, to prompt FP's topic theme*]
12 FBP: Some of them...married...But uh:: the one...Some I had by...
13 four...four...The four that graduated...
14 IBF: Uhhm
15 FBP: some I had... [*Repairs mark as she reaches for memory*]
16 I think –uh
17 IBF: um-hm
18 FBP: ...three or four graduated in (.4) Ball-timore:::
19 IBF: um-hm::
20 FBP: Douglass High School
21 IBF: Doug-lass:::

22 FBP Douglass:: (.6) High School ..(..8)
23 IBF: um-hm
24 FBP: ..in Baltimore (.4) Maryland.... (*pause*)
25 IBF: um-hm:: I just (.4) uhm:: (.4) came from Baltimore [=
26 FBP: =]You did::id
27 IBF: =]um-hmm...
28 I spent (.4) three months there[=
29 FBP: =] You did::: (.2) what did[=
30 IBF: =] um-hm (.4) I just
31 FBP: =] Really:: [=
32 IBF: =] I had to
33 FBP: =](.4) Where were (.2) you:::
34 IBF: =] I:: was (.2) O-kay (.2) I was (.4)
35 FBP: =]What street::[=
36 IBF: =] all (.2) OH-ver the place
37 FBP: =] all o-ver the place
38 IBF: yeah oh but (.2) I was::: (.4) What's street (.4) What's street (.4)::
39 What was the street:::
40 FBP: =]Is it Reisterstown Road...cause that's the bus-i-ness
41 IBF: um-hm...It was on the West side (.4) I think
42 FBP: =] West side? (.4) I was on...say the...I was on the north – west:::
43 IBF: uh-hm
44 FBP: =] That's on Reis-terstown Road:::
45 IBF: =] Old Frederick...Road:::
46 Old Frederick:: [*Refers to a major street in Southwest Baltimore on the way out to the suburbs*]
47 FBP: =]that's not too far...far as I can remember...right now:::
48 But I heard of other
49 IBF: =] um-hm
50 FBP: =[....kind of OUT...
51 IBF: =] um=hm
52 FBP: =] in the (.4) coun-ty
53 IBF: um-hm
54 I was out there...I was in Baltimore County....in Dun-dalk::
55 FBP: Yeah...that's the coun-ty

The female Black interviewer echoes the female Black Patient's topic opening of birth place with the cue "born and raised" in line 3 that prompts the beginning of a narrative form common in the Black community, referred to as "How I got ovuh" or tales of coming up and survival (Smitherman 1977). Both Lines 4 and 5 begin with "And..." to reflect the rhythm of how the story is being told. The female Black interviewer responds to the cue, picking up the topic of Baltimore, rather than returning back to the referenced "home," identified as Charlotte in line 1. In line 6, the Black female Interviewer uses a rapport-building interruption, intended to share a common experience of the city but does not gain a place on the conversational floor in line 9. Instead in lines 11, 14, 17 and 19, she uses a series of paralinguistic cues as prompts to encourage the female Black patient to speak about her children and their raising, preserving AD speaker autonomy in topic. Unlike many of the White–White encounters where a person with AD reaches for memory or difficulty with hearing loss is either corrected or marked by a topic change, the Black encounters in this set include more waiting and topic-related prompts, such as the repetition of the predominantly Black high school, Douglass, in line 21. The sounds at ends of words that are drawn out as prompts, marked by colons (:::) are common to holding the floor and prompts in Southern White and Black dialect, observed as extended syllables in encounters with Southern speakers (Wolfram *et al.* 1999). The use of extended single syllable words or last words in phrases were not observed in the European-origin English speaker in the reviewed recordings, establishing a suggestion of common patterns of Southern ethnicity recognized in participation and topics in common with the White female Interviewer and her clients.

From lines 26 to 34 in Segment 4, the phenomenon of a common collaborative floor appears once again with female–female interactions, characterized by speaker overlaps and rapport-building. The device appears across racial groups in this set of recordings as women build narratives with one another, recognizing common threads, though less so in the monoracial White interviews where professionals use information acquisition rather than exchange. In the Black–Black female encounter, the additional device of prosody or rhythm (identified with underlined words for emphasis) parallels what the speech partner emphasizes, a speaking resource in African-American Vernacular English (AAVE) considered vital to the meanings and significance of words being built together (Rickford & Rickford 2000). Perhaps a White speaker unfamiliar with Baltimore would not have acknowledged Douglass in quite the same way or with the same result. Following the "um-hm" prompts of recognition in lines 25 and 27, the Black female interviewer re-initiates her recent visit to Baltimore linked to common knowing. Where the Black patient refers to a somewhat segregated northwest Baltimore, the younger Black interviewer identifies she was "all over the place," especially the county suburbs that the [*older*] patient [*who had lived in more segregated times*] had only heard of "out there." Each recognizes implicitly what the other says about changes in segregation, cues that a White speaker without a common experience may have missed.

This story proceeds for over thirty-eight minutes, common to the Black–Black encounters where speech partners use recognized speech forms as resources to explore memories of the past as longer narratives, but less common in White–White encounters where professional agendas or lack of shared common experiences or indexical ground emerge from talk.

The more direct referencing of race occurs differently as well. In only one of the fifteen White-with-White speaker encounters was the topic of race referenced directly. The female White interviewer mentions that a particular day is a holiday, Martin Luther King Day, and the male White patient responds in a flat tone, "I don't consider that a holiday" and changes the subject. By contrast, Black patients speaking with the younger Black interviewer weave race throughout their narrative, in much the same way DuBois described as the double consciousness of living as a Black in a White society (DuBois 1953, 1961). The following Segment 5 captures one such narrative between a male Black Patient (MBP) and the female Black Interviewer (IBF):

Segment 5

1 IBF: C-E-P [*Is clarifying C's name*]
2 And this is our first conver-sa-tion over here
3 MBP: Yeah:::
4 IBF: ..But we been talkin' a lot all a-long:::
5 MBP: Well (.2) I tell yuh::::
6 I—I---I (.4) I know that my Grand-mother was bor- was SOLD:::
7 (.6) on the slave (.2) block (.6) in Rich-mond (.2) Virginia
8 (1.0) So evidently the-e people who bought her
9 carried her to South Carolina
10 And – uh:: (.4) There (.4) she was was on the Craig Plantation (.6)
11 C-R-E-I-G??
[*Clarifies the spelling for the listener*]
12 IBF: hmmm [*Agreement*]
13 MBP: Craig Plantation (.4) and he owned just as far as you could see
14 Any kind-a way you looked
15 IBF: hmm
16 MBP: So he had:: (.4) five or six (1.0) Nee-groe (.)
17 families and share-croppers
18 Now there were a couple of White familes there (.) too
19 but they use them as (.6)
20 su-per-visors (.4) or what-ev-er

21 But (.2) eventually they died out (.4) or what-ev-er
22 And:: (2.0) well my grand-mother was married to a – (.6)
23 to a fellow named Hen-ry
24 And you know (.2) they (1.5) take up the-e (.8)
25 plantation owner's last name:::
26 IBF: right::
27 MBP: ::: Once they freed the slaves... And he was a Craig (.4)
28 There-fore (.4) my grand mother was a Craig:::
29 IFF: um-hm
30 MBP And my grand- FA-ther was a Craig (.8)
31 And they did share-croppin'
32 Well s't's about as much as I know about my grand-mother
33 She must-a been born about 18- 50
34 Or maybe before that (.4) because they said when they freed the slaves
35 she was about thir-teen years old:::
36 IBF: [*Soft voice echoes*] °thirteen years old (.) hmm°
37 MBP: Maybe eight-teen forty-five (.4) or something like that
38 But any-way (.2) my::: (2.0) I think my grand-mother had
39 about two or three kids:::
40 No (.6) three? (.2) my grandmother had about.....
41 let's see (2.0) [*Soft voice follows*]
42 °two boys... and... two...°

In this interaction, the speakers take up memories from a previous encounter. The use of prosody marks their previous relationship and continuing narrative. In line 4, the use of *been* by the IBF functions as the non-recent perfective form in African-American English vernacular (AAVE) that carries inferred information (Labov 1998). Through the use of this resource, they are involved in a current talk still in progress, rather than a reminder of the past inferred by Whites who may hear the *have been* of a missing tense as Labov explains (1998: 135). Once again the use of paralinguistic prompts on lines 12, 15, 26, and 29 encourages the construction of a narrative of history and survival, rather than requests for direct information as if isolated from experience. The male Black patient holds the floor in the narrative through the use of rhythmic stressors of words as part of cadence, common to the female–female Black interactions as well, and through discourse devices such as listing and repetition observed across populations (Tannen 1989/96). No interruption follows the list or change of subject, but an affirmation in line 26 marks respect for the poetic construction. Marking differences in gendered practice, few interruptions or overlaps build topics in Segment 5, as seen in the more collaborative floors when women speak together. Just as shared conversational resources build the flow of a narrative, access to a common experience can build the substance of narratives.

For example, the socialization of young Blacks as members of certain communities or church groups may include interaction with elderly and ill persons as part of their religious or community service. For example, the visiting of the sick, elderly, and shut-ins forms an active part of the expectations of the young who attend the African Methodist Episcopal (AME) church (Campbell 1998). This intergenerational contact is often based on behavior expectations for the young person of sitting and listening respectfully to stories as part of being there for the older person. In this setting, it also shows regard for the patient's autonomy rather than the primacy of their own. This young Black interviewer seemed very skilled at creating a conversational floor and developing the narratives of the AD patients.

In the final Segment 6, a female Black patient tells a story about segregation.

Segment 6

1 IBF: So there was a lot of segregation..even when you were little?
2 FBP: Oh YEAH::::::: (..) Oh yeah. . . . It was ,,,, segregation after I MAR-Ried
[Sound stress signals it extended overran even longer period of her life]
3 They had–uh- the water fountains. . . . in the department stores and like. . .
4 Like umm. . . . it was [*Pause, long pause*] . . . time [*Pause*]
5 What was the name of those stores?:::. . . [*Voice trails off*]
6 [*Now lively again, storytelling*] ANY-way. . .
7 They had two water fountains. . . side by side. . .
8 IBF: um-hmm
9 FBP: They had COLORED over here and White over here. . . .
10 And that water was running up (..) from ../the same pipes\
11 IBF: [*Interrupts to echo in an overlap as if shared thought*]/same pipes\
12 FBP: They just separated when they get to those fountains
13 And I thought. . . I thought. . . . [*Switches slightly to different topic*]
14 And we had to ride in the BACK of the bus::::
15 Paid seven cents to ride the bus::
16 IBF: Seven cents?
17 FBP: Uh-huh . . . Even if there were . . . six . . . up front . . .
18 And we had to stand up anyway
19 Just in case Whites might come in . . . sit down . . . Blacks. . . .
20 We had to fill up from the back . . . Blacks..
21 The Blacks had to fill up from the back

22 And uh..Colored.....Bathrooms .. same way... Whites... Black...
23 And I remember my sister... She had married and::: uh..
24 went to Braddock Pennsylvania
25 When she came home [*Pause*] She went into the White [*Chuckles*]
26 ...she went into the White restroom
27 And they was lookin' at her... [*Chuckle in voice*] ..lookin' at her..
28 And She didn't say nothin about it...
29 She just went on in there and used the bathroom and came out [*Pause*]
30 But those White were looking,,.But they was looking at her..
31 But they didn't say nothing about it
32 IBF: um-hm (.4) Did she not know or not care?
33 FBP: She saw it up there... She saw it up there...
34 She just go into the bath- room
35 No matter..if there was the Colored or the White....
36 She just went on into the White bath room
37 You'd get to that before you'd get to the Colored...
38 Cause that was what they said..
39 Colored\
40 IBF: um-hm.... [*Murmurs*] colored
41 FBP: ..that's the way it was....
42 Colored people and White people.... Colored people and White people

The descriptions of discrimination occured frequently in the talk of Black patients with Black interviewer in the reviewed recordings. In these same race dyads, the subject is less likely to provoke denial or defensive response than with mixed-race dyads (Shipler 1997). The Black female interviewer, through prompts at transcript lines 8, 11, 16, and 32, provides encouragement and acceptance of the story without interruption. In similar situations, Whites may employ defensive strategies of aversion or denial (Gaertner & Dovidio 2000). In this narrative and others, the Black female interviewer positions the older person as the expert of their experience. Though also part of the life of southern Whites, stories of segregation appear less often in the narratives of those privileged by the system (Frankenberg 1993). For elderly Black patients, the denial or withholding of what may be a major theme such as segregation experiences would require constant editing and monitoring of one's talk. This barrier to communication would be a potential source of loss with a profound effect on therapeutic relationships in mixed-race dyads (Carter 1995). Any barriers to communication compound the increasingly difficult circumstances of social interaction as AD progresses over time.

Though interracial encounters were not available for this review, contrasts in permissible topics and speaking strategies of interaction that characterize the reviewed monoracial White and Black encounters suggest a need for more direct investigation of patterns of mixed-race dyads. Hecht, Collier, and Ribeau (1993) suggest that African-Americans identify seven issues as affecting their satisfaction in communication with European Americans. These are negative stereotyping, acceptance, expressiveness, authenticity, understanding, goal attainment, and powerlessness. Awareness of each of these needs to be included in any training for multicultural communication competence.

Conclusion

Based on our analysis of these samples of talk in context with other reviewed recordings, gender and ethnicity are clearly tied to social aspects of talk. In addition we noted that an institutional setting as well as the ages, social class, and the social hierarchy of professions of the participants appears to produce speaking practices, associated with social and cultural *habitus* (Bourdieu 1972, 1995). These practices, tied to our social identity, can function as either resources or sources of resistance in conversations. Those engaging either the elderly and/or persons with AD in conversation need to reflect on the consequences of particular speaking practices. Rather than mimic those with differing practices, health and human service workers may need additional information on the role of preferred speaking styles (Tannen 1993; Antaki & Widdicombe 2001) and what constitutes respect when social categories differ (Robinson 2002). They may need to explore ways of building participation (Hajek & Giles 2003) and to recognize, integrate, and promote variations that function as conversational resources. As health and human services professionals, it is our responsibility to accommodate social variations in talk when co-constructing conversations with persons who have AD. We can use our recognition of these speaking practices to not only improve our process of communication, but also to improve the process of our care.

References

Ainsworth-Vaughn, N. (1998) *Claiming Power in Doctor–Patient Talk*. NY: Oxford University Press.

Ambady, N., *et al.* (2002) "Physical therapists' nonverbal communication predicts geriatric patients' health outcomes." *Psychology and Aging* 17: 443–52.

Antaki, C. & S. Widdicombe, eds. (2001) *Style and Sociolinguistic Variation*. Cambridge: Cambridge University Press.

Bailey, B. (2000) "Communicative behavior and conflict between African American and Korean immigrant retailers in Los Angeles." *Discourse & Society* 11: 86–108.

Bethea, L. & A. Balazas (1997) "Improving generational health care communication." *Journal of Health Communication* 2: 129–37.

Bourdieu, P. (1972/1995) *Outline of a Theory of Practice*. Cambridge: Cambridge University Press.

Campbell, J. (1998) *Songs of Zion: The African Methodist Episcopal Church in the United States and South Africa*. Chapel Hill, North Carolina: University of North Carolina Press.

Carter, R. (1995) *The Influence of Race and Racial Identity on Psychotherapy. Toward a Racially Inclusive Model*. NY: John Wiley & Sons-Interscience.

Coates, J. (1997) "The Construction of a Collaborative Floor in Women's Friendly Talk." In *Conversation: Cognitive, Communicative, and Social Perspectives*, ed. T. Givon. Amsterdam/Philadelphia: John Benjamins.: 55–89.

Cooper, L. & D. Roter (2003) "Patient–Provider Communication: The Effect of Race and Ethnicity on the Process and Outcomes of Healthcare." In *Unequal Treatment: Confronting Racial and Ethnic Disparities in Healthcare*, eds. B. Smedley, A. Stith, & A. Nelson. Washington, DC: National Academies Press: 552–625.

Coupland, N., *et al.* (1991) *Language, Society, & the Elderly*. Oxford: Blackwell Publications.

Day, D. (1998) "Being Ascribed, and Resisting, Membership of an Ethnic Group." In *Identities in Talk*. C. Antaki & S. Widdicombe. London: Sage: 151–70.

Derber, C. (2000) *The Pursuit of Attention: Power and Ego in Everyday Life*. Oxford: Oxford University Press.

DuBois, W.E.B. (1953/1961) *The Souls of Black Folk*. Greenwich, CN: Fawcett.

Duranti, A. (1997) *Linguistic Anthropology*. Cambridge: Cambridge University Press.

Ellis, D. (1999) *Crafting Society. Ethnicity, Class, and Communication Theory*. Mahwah, NJ: Lawrence Erlbaum.

Emery, V.O. (1999) "On the Relationship between Memory and Language in the Dementia Spectrum of Depression, Alzheimer Syndrome, and Normal Aging." In *Language and Communication in Old Age: Multidisciplinary Perspectives*, ed. H. Hamilton. NY: Garland.: 25–62.

Ervin-Tripp, S. (2001) "Variety, Style-Shifting, and Ideology." In *Style and Sociolinguistic Variation*, ed. J. Rickford. Cambridge: Cambridge University Press: 44–56.

Ervin-Tripp, S. & A. Kuntay (1997) "The Occasioning and Structure of Conversational Stories." In *Conversation: Cognitive, Communicative, and Social Perspectives*, ed. T. Givon. Amsterdam and Philadelphia: John Benjamins: 133–66.

Foddy, M. & Y. Kashima (2002) *Self and Identity: Personal, Social, and Symbolic*. Mahwah, NJ: Lawrence Erlbaum.

Foucault, M. (1977/1991) *Discipline and Punish. The Birth of the Prison*. NY: Vintage Books.

Frankenberg, R. (1993) *The Social Construction of Whiteness: White Women, Race Matters*. Minneapolis, MN: University of Minnesota Press.

Gaertner, S. & J. Dovidio (2000) *Reducing Ingroup Bias: The Common Ingroup Identity Model*. Philadelphia, PA: The Psychology Press.

Goldberg, J. (1990) "Interrupting the discourse on interruptions: An analysis of relationally neutral, power- and rapport-oriented acts." *Journal of Pragmatics* 14: 883–933.

Hajek, C. & H. Giles (2003) "New Directions in Intercultural Communication Competence: The Process Model." In *Handbook of Communication and Social Interaction Skills*, ed. J. Greene & B. Burleson. Mahwah, NJ: Lawrence Erlbaum.

Hanks, W. (1996) *Language and Communicative Practices.* Boulder, CO: Westview Press.

Hecht, M., *et al.* eds. (1993) *African American Communication: Ethnic Identity and Cultural Interpretation.* Newbury Park, CA: Sage.

Holtgraves, T. (2002) *Language as Social Action. Social Psychology and Language Use.* Mahwah, NJ: Lawrence Erlbaum.

Hopper, R. & C. LeBaron (1998) "How gender creeps into talk." *Research on Language and Social Interaction* 31: 59–74.

Housely, W. & R. Fitzgerald (2002a) "The reconsidered model of membership categorization analysis." *Qualitative Research* 2: 59–83.

Howard, D., *et al.* (2002) "Distribution of African Americans in residential care/ assisted living and nursing homes: More evidence of racial disparity?" *American Journal of Public Health* 9: 1272–7.

Itakura, H. (2001) "Describing conversational dominance." *Journal of Pragmatics* 33: 1859–80.

Jacoby, S. & E. Ochs (1995). "Co-construction: An introduction." *Research on Language and Social Interaction* 28: 171–83.

Koch, L., *et al.* (2001) "Attitudes toward Black English and code switching." *Journal of Black Psychology* 27: 29–42.

Labov, W. (1998) "Co-existent Systems in African American Vernacular English." In *African-American English: Structure, History, Use*, eds. S. Mufwene, J. Rickford, G. Bailey, & J. Baugh. London: Routledge: 110–53.

Labov, W. (2001) "The Anatomy of Style-Shifting." In *Style and Sociolinguistic Variation.* P. Eckert & J. Rickford, eds. Cambridge: Cambridge University Press: 65–108.

Linell, P. (1998) *Approaching Dialogue: Talk, Interaction, and Contexts in Dialogical Perspectives.* Amsterdam/Philadelphia: John Benjamins.

McDougall, H. (1993) *Black Baltimore: A New Theory of Community.* Philadelphia, PA: Temple University Press.

Miller, P., *et al.* (1990) "Narrative practices and the social construction of self in childhood." *American Ethnologist* 17: 292–311.

Morgan, M. (1996) "Conversational Signifying: Grammar and Indirectness among African American Women." In *Interaction and Grammar*, eds. Elinor Ochs, Emanuel A. Schegloff, Sandra A. Thompson. Cambridge: Cambridge University Press: 405–34.

Ochs, E. (1992) "Indexing Gender." In *Rethinking Context: Language as an Interactive Phenomenon.* A. Duranti & C. Goodwin, eds. Cambridge: Cambridge University Press: 335–58.

Omi, M. & H. Winant (1996) *Racial Formation in the United States. From the 1960s to the 1990s.* NY: Routledge.

Paoletti, I. (1998) "Handling 'Incoherence' According to the Speaker's On-Sight Categorization." In *Identities in Talk*, eds. C. Antaki & S. Widdicombe. London: Sage, 171–90.

Papousek, M. (1995) "Origins of Reciprocity and Mutuality in Prelinguistic Parent and Infant 'Dialogues'." In *Mutualities in Dialogue*, ed. K. Foppa. Cambridge: Cambridge University Press: 58–81.

Reichman, J. (1997) "Language-specific response patterns and subjective assessment of health: A sociolinguistic analysis." *Hispanic Journal of Behavioral Sciences* 19: 353–68.

Rickford, J. (1999) *African American Vernacular English. Features, Evolution, Educational Implications.* Oxford: Blackwell.

Rickford, J. & R. Rickford (2000) *Spoken Soul. The Story of Black English.* NY: John Wiley.

Ripich, D., *et al.* (2000) "Conversational cohesion patterns in men and women with Alzheimer's disease: A longitudinal study." *International Journal of Language and Communicative Disorders* 35: 49–64.

Robinson, M. (2002) *Communication and Health in Multi-Ethnic Society.* Bristol, UK: Policy Press.

Roter, D., *et al.* (1991) "Sex differences in patients' and physicians' primary care medical visits." *Medical Care* 29: 1083–93.

Saunders, P. (1999) "Gossip in an Older Women's Support Group: A Linguistic Analysis." In *Language and Communication in Old Age: Multidisciplinary Perspectives.* H. Hamilton, ed. NY: Garland: 267–93.

Schegloff, E. (1991) "Reflections on Talk and Social Structure." In *Talk and Social Structure: Studies in Ethnomethodology and Conversation Analysis*, eds. D. Boden & D. Zimmerman. Berkeley, CA: University of California Press: 44–70.

Shipler, D. (1997) *A Country of Strangers: Blacks and Whites in America.* NY: Alfred A. Knopf.

Smitherman, G. (1977) *Talkin and Testifyin. The Language of Black America.* Detroit, MI: Wayne State University Press.

Street, R.J. (2003) "Interpersonal Communication Skills in Health Care Contexts." In *Handbook of Communication and Social Interaction Skills*, eds. J. Greene & B. Burleson. Mahwah, NJ: Lawrence Erlbaum: 909–33.

Stubbe, M. (1998) "Are you listening? Cultural influences on the use of supportive verbal feedback in conversation." *Journal of Pragmatics* 29: 257–89.

Tannen, D. (1989/1996) *Repetition, Dialogue, and Imagery in Conversational Discourse.* Cambridge: Cambridge University Press.

Tannen, D., ed. (1993) *Gender and Conversation Interaction.* NY: Oxford University Press.

Tannen, D. (1993) "The Relativity of Linguistic Strategies: Rethinking Power and Solidarity in Gender and Dominance." In *Gender and Conversational Interaction*, ed. D. Tannen. NY: Oxford University Press: 165–88.

Terrell, B. & D. Ripich (1989) "Discourse competence as a variable in intervention." *Seminars in Speech and Language* 10: 282–97.

Thimm, C., *et al.* (1995) "Power-related talk: Control in verbal interaction." *Journal of Language and Social Psychology* 14: 382–407.

van den Brink-Muisnen, A., *et al.* (2002) "Do gender-dyads have different communication patterns? A comparative study in Western-European general practices." *Patient Education and Counseling* 48: 253–64.

Weber, M. (1947) *The Theory of Social and Economic Organization.* NY: Free Press.

Williams, A. & J. Nussbaum (2001) *Intergenerational Communication across the Life Span*. Mahwah, NJ: Lawrence Erlbaum.

Wolfram, W., *et al.* (1999) *Dialects in Schools and Communities*. Mahwah, NJ: Lawrence Erlbaum.

Zimmerman, D. (1998) "Identity, Context, and Interaction." In *Identities in Talk*, eds. C. Antaki & S. Widdicombe. London: Sage: 87–106.

The authors wish to acknowledge the assistance of Ms. Naomi Sampson and Ms. Tiffani Myers in the preparation of this chapter.

4
Talking in the Here and Now: Reference and Politeness in Alzheimer Conversation

Boyd H. Davis and Cynthia Bernstein

Background

We first met "Larry Wilcox" in December, 1999, and learned something of him throughout 2000 and 2001, the two years during which we conversed with him, and on which we base this case study of reference and politeness in Alzheimer conversation. Some of our conversations were short and some were shorter – Wilcox was not loquacious. Indeed, staff in the Alzheimer's unit at Pleasant Meadows, a private retirement and assisted living facility in Charlotte, NC, had initially wondered whether he could be sufficiently talkative with strangers to be a conversation partner. We were told that he was in his early 80s, diagnosed with probable Alzheimer's Disease, no longer formally assessed, but in the process of moving from moderately severe to severe cognitive decline. We attempted to get a baseline for cognition by administering the Seven-Minute Screen (Solomon *et al.* 1998), but he stopped the test two-thirds of the way through, and refused for the rest of our acquaintance to participate in any interaction where the conversation partner carried notebooks or picture cards or asked content-seeking questions.

Instead of administering any kind of formal testing, which he and others in the unit were largely unwilling to undergo, we looked at activities of daily living. Wilcox increasingly needed assistance across 2000–2001 with choosing clothes, dressing and bathing. On occasion, he proffered fragments of facts and events about himself and others in his family, his work, and his life, but from much earlier times of his life. His ability to handle the functions of daily life matched descriptions of categories Five and Six on the Global Deterioration Scale for Dementia, suggesting his move from moderate to moderately severe dementia during that time (Reisberg *et al.* 1982; Reisberg 1988). In 2002, he spoke but little, sleeping or dozing most of the day. He died early in 2003.

Social politeness

When linguists discuss the ways people show politeness in conversation, they are often concerned with how different facets of the communicative interaction, from its setting, situation, speakers and sequence to its genre and expected norms or conventions (Hymes 1972), can be seen to index underlying issues of power, status, and hierarchy (Brown & Levinson 1987; cf. Fraser 1990). They could be concerned with how the speakers draw on strategies of involvement with each other to create solidarity, or on strategies of independence, perhaps to avoid placing each other in uncomfortable situations (Scollon & Scollon 2002). Analysts can focus on conversational acts that could be a threat to the "face" of either speaker (Gumperz 1982a; Brown & Levinson 1987) or a threat to the transactional goal of the interaction (Merrison 2002). In our review of 41 jointly constructed conversations archived over during the two years, we have moved to assume an interpretation of politeness keyed first to discussions by Sabat and his co-authors over a decade (Sabat 1992; 2002; Sabat & Harré 1992; Sabat & Cagigas 1997; Temple *et al.* 1999) and then to a distinction between tact as facework on the one hand and social politeness as a relational goal on the other hand, by Rhys and Schmidt-Renfree (2000). For this discussion, we will look at interconnections between referentiality and the relational aspects of social politeness, and review how rapport was co-constructed, in ways very similar to those described by Davies in her "situated interpretation of joking as a speech activity" constructed by native and non-native speakers (Davies 2003: 1361).

By referentiality, we mean how speakers offer verbal landmarks and pathways that allow listeners to follow what is being talked about, whether it is direct, such as providing the name of someone in one clause (*Bobby ate the cheese*) which explains the pronoun in the next clause (*and he liked it*), or indirect, in which the listener infers the reference in some other way (Ted: *Is something wrong?* Jill: *Hey, didn't you see the ambulance in the driveway?*). As suggested by the exchange between Ted and Jill, social and cultural features are often needed to explain referentiality. In this (oversimplified) example, ambulances are assumed to be vehicles which are asked to move from public, medically affiliated areas into private space in order to retrieve people who are sick or injured, via driveways which abut housing or office space. Since a sighting of an ambulance indexes a nearby problem, we assume that Jill expects Ted to have seen the ambulance and to have recognized that something was happening which was neither happy nor usual, and that she also expects him to have the conversational competence to recognize her expectations.

Speakers with Alzheimer's vary in their competence, including their ability to use reference, and they vary greatly enough to underscore the importance of using spontaneous or minimally prompted conversation in addition to or as supplementary to more formal clinical testing. For example, one of our speakers in the Alzheimer's corpus shifts ordinary reference to manage topic, while another speaker can use only context-driven referentiality, in an effort to continue conversing. A focus on the local situation, and the local context for interpretation, supports the move to analyze the social and cultural features of reference in conversation (Schiffrin 1994: 407). What Hengst notes as a change in perspectives on aphasia research has begun to appear as well for research on Alzheimer's. Hengst comments:

> In the last decade, many researchers and clinicians, taking up interactional sociolinguistic perspectives on communication and aphasia, have shifted away from controlled tasks and begun to focus on the communicative competencies of individuals with aphasia in conversations and everyday activities.... Interactional perspectives on language and communication challenge each of the conventional tenets about referencing.... (Hengst 2003: 831–2)

Perkins *et al.* (1998) claim that spontaneous speech and minimally prompted conversation in normal social settings are far more natural and furnish lexical and pragmatic data too valuable to ignore, whether for diagnostic purposes or for the purposes of describing DAT speech in order to sustain the social identity and the quality of life of the speaker. They note that a very real problem for researchers is the impact of the deficit model and the over-emphasis on aggregational analysis: "there is a vast amount of individual variability within diagnostic categories that is potentially obscured by presentation of group data if that is all one plans to present. Group data, while undeniably useful, has a way of blinding practitioners to the need to look more closely at the data and at the individual". Sampling conversation, they claim, is the "most representative, most ecologically valid, and least artificial" way of eliciting language to be studied. Conversation was at the heart of the work by both Hamilton (1994a) and Ramanathan (1997) as they drew on multi-year conversations with Alzheimer's speakers to ground qualitative discussions of Alzheimer's narrative and conversational discourse.

As part of her ground-breaking study of conversations with "Elsie," an Alzheimer's speaker, Hamilton detailed how Elsie's decrease in her ability to "take the role of the other" was not uniform. We find both of our speakers to pattern in similar ways, and underscore Hamilton's use of a functional approach and Halliday's (1978) terms: *mode*, for procedural interactions, *tenor*, for managing interpersonal positioning, and *ideation*, or content. Although Elsie showed problems with ideation (such as with pronouns) in her earliest conversations, she successfully maintained tenor, as she presented accounts and positive politeness the following year, and was still able to handle mode, or turn-taking and responding to questions, even in her latest conversations (Hamilton 1994a, 41–2; cf. 1994b). Ramanathan (1997) demonstrated that interaction across settings was as important as the content of the conversation, if not more so, in terms of the Alzheimer's speaker's participation. Temple *et al.* (1999) studied the retention of politeness, finding that "traditional clinical methods and psychometric testing techniques [...] may produce an incomplete, and perhaps distorted, picture of the Alzheimer's Disease sufferer" while "enhanced social contexts and specialized communication techniques may work to increase their level of social functioning and communication" (164f.). Rhys & Schmidt-Renfree proposed that politeness be distinguished from tact, in order to distinguish facework from other kinds of social strategies that are polite and conventional, even formulaic (2000: 542) routines of conversation, such as greetings or compliments (2000: 536). They suggested, following Sabat and others, that while an Alzheimer's speaker may not always be able to manage facework concerns of their partner, they are keenly aware of their own face needs. Referentiality is one of the aspects of conversation in interaction that comes into play at this point.

Our case study of Larry Wilcox suggests that part of the repertoire for referentiality that DAT speakers retain is tied to what Ripich *et al.* call

> their compensatory pragmatic aspects ... Persons with AD appear to strive for communication competence as language declines by increasing certain compensatory pragmatic aspects ... at different severity levels of the disorder. This suggests that compensatory pragmatic devices are used with flexibility, and that the desire to communicate is maintained throughout the course of AD (2000: 217).

Following Sabat (2002), we see this effort to compensate as part of a desire to present the self in positive ways and to have that Self recognized.

In his discussion of interaction with Mrs. F., a conversation partner with Alzheimer's, Sabat demonstrates how she indicates an awareness of her own social limitations. Accordingly, we have reviewed interactions with Wilcox for signals of awareness of social limitations within the discourse, or interactive limitations, seeing those signals as part of the referentiality demanded by conversation in interaction. Referentiality in this sense goes beyond saying what a word means, or what a pronoun might refer to. It is instead a referentiality that includes structural elements such as an awareness of discourse roles and of macropropositions (Kintsch & Van Dijk 1978) for the conversation, including the fact that a conversation is going on and that each of the conversants is aware of being included.

A sense of dialogic inclusion is at the heart of much of Tannen's analyses of conversational style. For example, her summary of schemata for politeness as language used in "conventionalized ways of honoring those needs . . . to be involved with others and to be left alone," is in dialogue with Durkheim's (1915) distinction between negative and positive religious rites and rituals, which Goffman (1967) in turn drew upon to explain the notion of deference. Deference is explained as appreciation, expressed either in avoidance rituals that have rules about privacy and respect, or as presentational rituals, which include "salutations, invitations, compliments" (Tannen 1999: 462). It is these rituals that remind the recipient that s/he is not and may not be an island. And it is those rituals which the Alzheimer's speaker retains, using them for social, relational goals for conversational interaction when transactional goals for the conveyance of specific information can no longer be attained.

Referentiality and interaction: fine-tuning functions of *thing*

In 1991, Kempler summarized findings to date of pragmatic deficits in the communication by Alzheimer's speakers, including their being "grossly insensitive to the needs of the listener" particularly with regard to topic management, and their presenting a decrease both in informational content and in coherence. Ulatowska had noted in 1988 that "decreased informativeness could be traced back, at least in part, to problems in reference such as the overuse of demonstratives . . . and exophoric reference (e.g., this without a clear antecedent)" (1991: 104). In the late 1980s, Ripich and Terrell had identified "referential errors as the most frequent coherence error;" while Hier *et al.* found Alzheimer's reference to be characterized by an overuse of pronouns (meaning without antecedents) and "empty" words (Ulatowska 1991: 116–17).

Shadden's (1998) summary of verbal fragmentation in Alzheimer's talk includes the categories of "empty speech" listed in 1985 by Nicholas *et al.* (Shadden 1998: 53) In addition to neologisms, Shadden notes that several types of paraphrasias (half-right words, word substitutions), repetition, and the replacement of logical conjunctions such as *so* and *because* by *and*, all contribute to an assessment of speech as being empty of meaning, and cites Nicholas *et al.* for categories such as empty phrases used as continuation devices (*and so on, like that*), indefinite terms (*thing*), deictic terms (*that*), pronouns with no antecedents, and comments on a particular task (*It's hard for me to say that*).

More recently, researchers have identified additional or different functions for several of these categories. Just as the frequency of appearance of a particular form such as *that* may not be as important as its range of functions, the social-interactive function of a word such as *thing* used as a discourse extender (in *things like that*) may be as important to a DAT speaker for building and maintaining social relationships as it apparently is to adolescents. A recent study by Overstreet demonstrates that extenders are more than vague expressions; they are used "to indicate assumptions about shared knowledge and experience, or to mark an attitude toward the message expressed, or toward the hearer" (Overstreet 1999: 11; cf. Overstreet & Yule 2000). Almor *et al.* (1999) show "that AD patients' pronoun problems do not necessarily imply a discourse impairment but instead could be seen as the outcome of normal discourse processing in the context of reduced computational ability" (and cf. Almor *et al.* 1999). "*Thing*" may be a noun whose use as a pro-form increases with the speaker's age; however, it may be that words like "*stuff*" and "*thing*", through a social-relational function such as phrasal extension, contribute to the transactional goals of an interaction by being relational, in that they present opportunity for alignment and the establishment of shared knowledge, no matter how tenuous or how temporary.

To get some sense of the varied functions for *thing* among aging adults in general, we analyzed conversation and interviews of regional speakers from the Charlotte Narrative and Conversation Collection and the Oral History Project, both of which are held in UNC–Charlotte's digital New South Voices project (http://www.newsouthvoices.uncc.edu). We concorded conversations with eight men and 16 women in Wilcox's age cohort (65–84), 3 whose conversation was recorded 21 years earlier when they were the same age, and 9 from the cohort a generation younger (45–64). By looking at the functions for "*thing,*" we can see better how to begin to assess changes in usage over time, keyed not to frequency of appearance but to its range of usage or functions.

Table 4.1 Initial tabulation of functions for *"thing"* from the Charlotte Narrative and Conversation Collection

function	*example*
direct object ["patient"] pro-form typically substituting for • recipient of action • event/situation • discrete, countable object • abstraction or mental activity	I thought we'd have time to explore and do things and one Christmas, things were really difficult (broke) I was going to sell the things in Mother's house my, that thing hurt, honey [thing = heartbreak]
fronted/anticipatory pro-form/phrase	The thing is, we . .; the only thing I could say was . . .
extender phrase, used in interaction, for solidarity: Overstreet 1999;	and all that kind of/sort of thing (see clichéd phrase)
emphasizer-evaluator phrase, used in interaction, with narrative embedded in the interaction	and of all things I had my tonsils out; that was the surprising thing, you know
General: clichéd phrase [for lexical bundle, see Biber *et al.* 1999; cf. extender]	(all) that sort of thing, and of all things
pro-form with appositional force within phrase: adds colloquial flavor, may be solidarity-builder	she jus' the meanest thing you ever saw
lexical: reduce humanness	no, I think that was some, some thing that we saw
lexical: sexual euphemism	well, it was his thing [thing = penis]

When one adds forms such as *anything* or *something*, the range of functions is extended to include ambiguity and the presentation of deliberately indefinite reference: for example, "No, I think *something* spooked that pony" includes an element of the counter-factual.

Our data suggests that the use of formulaic phrases, and what Biber *et al.* call lexical bundles (Biber *et al.* 1999) is keyed to individual preferences and to perceived social situations. *Thing*-phrases occur either at utterance initial, where they front the information and change the information structure of the conversation, or utterance-final, where they typically act as discourse extenders and seem strongly tied to social-interaction functions. The oldest cohort, women taped in 1979, were talking with animation about topics they had chosen, but their conversation partners were strangers: their lexical bundles were fronted ("The *thing* I want to tell you is"); they used no discourse extenders with "*thing*" (though they did extend discourse via prosody), and they typically did not use the pro-form. The same-age cohort presented the full range, as did the younger-generation cohort: in each group, every individual presented proforms, formulaic phrases as extenders, phrases and formulaic phrases as emphasizers and evaluators, and fronted pro-forms and phrases, though to different degrees.

We first reviewed changes in the use of "*thing*" and its compounds in the conversations by Ms. Copeland, who has mild Alzheimer's. In 2000, her fluent conversation and self-initiated narratives presented pro-forms, clichéd or formulaic phrases as discourse extenders, and the use of phrases with "thing" to emphasize and evaluate actions or events of a narrative she was sharing. Over 2001–3, and in mid-2004, she has continued to present pro-forms, but her clichéd phrases now typically occur utterance-initial. It is as if she were buying a fraction of time to think, retaining the floor as a means of maintaining social connection. We did not meet Wilcox until his disease was much further advanced, when he could present only fragments of narrative in collaborative conversation (cf. Goodwin 1995; 2003 on co-constructing conversation with an aphasic man).

Wilcox used the word "*thing*" in only two ways: to refer to a locatable object that he and we were jointly observing, as in 1(a) and 1(b) or in a colloquial phrase which suggested humorous mitigation of a negative comment, as in 1(c).

(1a): in this excerpt, "*thing*" refers to Wilcox's nose.

BD: Linda, have you had a chance to meet LW?
LM: I did some time ago. Hi, Larry.
LW: Hey, hey.
LM: My name is Linda. Do you have a cold today?
LW: I sure do – that *thing*'s running a mile a minute.

(1b): in this excerpt, "*thing*" refers to the portable tape recorder being used to record conversations.

BD: You hold this while I get me a chair, would you please?
LW: I will, OK. ==Good gracious==.
BD: ==Okay==
LW: That *thing's* heavy.
BD: Isn't it, though!

(1c): in this excerpt, "*thing*" is part of a colloquial and formulaic phrase often used for emphasis or humorous mitigation; the phrase itself occurs appropriately

LW: I was born in Yadkin County
LM: Yadkin County. Okay..
LW: There ain't no such *thing* as that now, not up my way.
LM: Not up your way?
BD: Yadkin County changed.
LW: Yeah.
BD: Yeah.

When Wilcox used formulaic, frozen phrases or regional colloquialisms as in (1a): *running a mile a minute*) or (1c): *up my way, ain't no such thing*), he shifted both his prosody and his pronunciation (Davis *et al.* 2000; Guendouzi & Müller 2001). More typically, Wilcox used *some/anything*, matching two of the functions displayed by "*thing*" (and for *some/anything*) by the same-age range, non-impaired persons living in the same region. Again, these uses were as a substitutable pro-form and as part of a phrase used for emphasis and evaluation, both functions restricted to the here-and-now. Pro-forms could substitute only for a discrete, countable, and currently visible object (2a), and an event or situation or an abstraction or mental activity mentioned in the previous conversational turn (2b). Phrases for emphasis or evaluation (2c) were colloquial and formulaic (in the sense of Wray & Perkins 2000; Wray 2002).

(2a) We have asked Wilcox to circle an item on a form and have just handed him a piece of paper

LW: I need that sitting on *something*

(2b) We have just commented on the renovation occurring in his wing of his unit:

LW: ==A few==little minor touches
BD: Yeah, like what?
LW: Well, just like – somebody picks us each of us new paint,
BD: Oh?
LW: Just, just enough to be noticeable
BD: Mmhum mmhum
LW: It's *something* that's not even needed right now but they can always get it just the same

(2c) We have just asked Wilcox what he wants to fish for:

LW: *Anything* that bite the hook!

The examples we have cited suggest that "empty words" for Wilcox are usually not truly empty, because they often have social-relational function. He maintains indefinite adjectives and pronouns, extender-phrases, and substitute forms such as "thing," a finding which may be worth examining in detail for other AD-speakers. Understanding more about the range of functions for indefinite constructions in addition to their frequency of occurrence may be useful in identifying where and how the Alzheimer's speaker is attempting to convey both social relationship and information. What is confusing for the interlocutor – to the extent of blocking joint construction of meaning – is Wilcox's use of third-person personal pronouns with neither given nor identifiable referents (Fox & Thompson 1990), particularly in conjunction with *"thing"*.

In addition, the data illustrate how Wilcox draws on his remaining grammatical resources to support continued social interaction, and social politeness, in at least two ways. First, he continues to offer emphasis and evaluation, if usually through well-worn formulae, in appropriate parts of the turn sequence and at expected positions in the sentence. Then, he never refuses to respond to questions in those conversational turns that ask him about something in the here and now, or are in the immediately preceding clause, which means he can offer some kind of answer to the last question in a series. And, in institutional settings, until a conversation partner and the speaker with Alzheimer's can either identify or evolve small-talk routines beyond the initial greeting, a good bit of their "conversation" is enacted through question–answer sequences, no matter how awkward (see Ramanathan-Abbott 1994; 1997 on the impact on interaction of changing settings; see Shakespeare 1998 on sustaining conversation with confused speakers).

That: social interaction via a cohesive marker with demonstrative reference

The agreement of a pronoun with its preceding referent is part of a speaker's cohesion. Cohesion analysis is, as Liles and Coelho note, "useful in the characterization of language-impaired individuals' narrative texts in terms of their ability to organize content" (Liles & Coelho 1998: 65; cf. Davis & Coelho 2004), and underlies components in assessments of pragmatic skills needed for conversation (for several such assessments, see Coelho in Cherney *et al.* 1998). We reviewed all demonstratives in Wilcox's discourse, focusing on the functions of "*that*," finding some changes in pronominal usage over the two-year period.

Wilcox retains both adjectival and pronominal usage of *that* in both 2000 and 2001, with his adjectival use of *that* almost invariably with reference to nearby (proximal), visible, and locatable items. For example:

(3a) *that* thing's running a mile a minute (2000: "*that* thing" = nose)
boy I appreciate *that* water (2000: water he has just drunk, which interlocutor brought)
already had all of *that* stuff (2001: "*that* stuff" = cancer)

That-initial is frequently used turn-second either to align with what the interlocutor has just said, or to continue positive affect, in both 2000 and 2001:

(3b) *that* would be fine
that ought to clear it up for us
that's my suggestion
that's all right with me
that's nice of you
that's about what mine are for

Reference may be either locatable/visible or inferable/identifiable. In (3a), initial-*that* signals "I'm the one who did it and I implicitly accept the compliment":

(4a) LM: Somebody combed your hair awfully pretty!
LW: *That* was me! (2001)

In (4b), his first *that* is cataphoric and his second *that* was interpreted at the time as possibly referring to the main entrance where both the facility van and local bus stop; roses and other flowers are planted nearby. Wilcox does not contradict LM's replacement of "rosebuds" for his "rosewoods" and a few phrases later, responding to her prompt, he mentions his mother's roses. After a long pause, Wilcox says

(4b) LW: I want to go and see how they're doing before I leave.
LM: Oh, OK. Before you leave, huh? Where will you be going?
LW: They got a lot of, they got a lot of rosewoods on the stop.
LM: A lot of rosebuds on the stop?
LW: Yeah.
LM: Um...

LW: *That*'s going to be awfully pretty when they open *that* up!
LM: Oh, absolutely!
LW: Yeah.
LM: You like rosebuds, too?
LW: *That*'s my favorite!

****

LM: Did your Mama plant rose bushes?
LW: My mother had awfully pretty ones this year.
LM: She did? This year?
LW: She won't let me get in them! She's afraid I'll trim it up!

That-initial also initiates formulaic phrases expressing frustration or other negative affect:

(5a) *that* wouldn't help me, *that*'s the catch, *that*'s the way it goes, OK. *That*'s, *that*'s enough

and occasionally announces identification of a locatable, visible object or person.

(5b) *that*'s mine, *that*'s a watch, *that*'s Mrs. Kennedy, I believe, *that* was the girl I saw

Pronominal-*that* in clause-final position occurs exclusively in 2000 in his Seven-Minute Cognitive Exam; the first in the list of examples refers to taking the exam itself; the others are his comments after viewing a picture-card, offering an incorrect answer [such as "banana" for a card with a picture of a bunch of grapes], and learning the correct answer. A discussion of *that*-final follows below.

(6a) Well, I'll try *that*, I didn't guess *that*, I don't even remember *that*, What the hell is *that*?
I like *that*: now you tell me what it is.

That-medial can function either as a relative pronoun (*You had birds that liked flesh?* in response to the interlocutor's comment about a bird that had pecked her) or may play conventional clausal roles. Its reference is to something jointly locatable and directly observable, or to the last clause in the immediately preceding turn.

(6b) It's something *that*'s not needed (2000: 2b above, referring to renovations)
I reckon *that*'s so (2000: agreeing with interlocutor comment about weather)
I wasn't studying *that* then (2000: response to learning correct answer to previous question)
They sure kept *that* hid while I was there! (2001: response to correct name for face card)

However, in 2001, *that*-referentiality in several phrases with *that*-medial, was non-phoric in the sense *that* the pronouns appear referential but actually have no referent at all except in the speaker's mind (Strauss 2002: 145; Halliday & Hasan 1976: 61).

Indeed, much of Wilcox's medial- and final-*that* usage in 2001 suggests that interpreting *that*-referentiality throughout 2001 depends to some extent on the position of *that* in the phrase, and, to a lesser extent, whether its use is preceded by referentless pronouns *she* and *they*. Final-*that*s typically occur as formulaic phrases serving as turn-second responses, though often not clearly tied to (or identifiable with) the previous statement. These phrases are more probably phatic, inserted where it would be polite to show alignment with the interlocutor, and usually said with exclamatory speech contour:

(7a) Been trying to do *that*! | I don't do *that*!
You can tell *that*! | I wouldn't say *that*
I best forget *that*! | She's got *that* right!

In his 2001 conversations with one interlocutor, GN, Wilcox offers fewer referents which are textually locatable or inferable/identifiable, although this was not the case in 2000. Hoping to initiate conversation by presenting a topic Wilcox would take up and co-construct, GN often offers him multiple cues as serial questions in a single turn, and frequently brings a memory-prop such as a calendar. And as discussed, Wilcox typically responded to the last question or the last clause in such a series. However, over 2001, Wilcox's formulaic phrases of response, while syntactically correct, and inserted in appropriate positions in the utterance, offered progressively less context that GN can identify and from which he could create additional prompts or extendable remarks during the conversation. In Appendix A, we present a typical conversational interaction between the two, including a series of topics to which Wilcox does not respond, and what is clearly Wilcox's effort to shift topic and to achieve relational goals of maintaining social politeness, and continuing interaction, when Wilcox stops the current topic with an offer: "Come slide up and sit down so you don't have to strain yourself *like that*."

To see if Wilcox's demonstrative pronouns presented a distinction between proximal and distal, or seemed instead to be working from another set of referential distinctions, we also examined his usage of "*this*," following Strauss (2002). Strauss maintains that the distal–proximal model for demonstrative reference is static and not particularly reflective of what occurs in conversational interaction, so that a dynamic model based on deixis is more useful. Her model incorporates "the relative importance of the referent being marked by a demonstrative, the relative newness or givenness of that referent, and most importantly, the relationship between the interlocutors and the role of the hearer" (Strauss 2002: 133). Based on a review of deixis, Strauss presents a model keyed to focus instead of to proximity, where focus is seen as the variable "degree of attention the hearer should pay to the referent," that is, varying across high, middle and low, with the implication being that

> the speaker frames the referent as shared, unshared, or somewhere in between, or as important, unimportant, or somewhere in between to accomplish certain interactional goals and to express certain attitudinal stances vis-à-vis the referent in question or vis-à-vis the interlocutor (2000: 135).

Strauss draws on a corpus of 45,000 words and 41 interactions (TV news interviews, radio talk show, university lecture-discussion, and multi-party talk among friends) to claim the following: In general, *this* has high focus; *that* has intermediate; and *it* has low focus because *this* co-occurs with new information, while *that* and *it*, being primarily anaphoric, co-occur with given or shared information, suggesting shared understanding. *That*, whether exophoric or non-phoric, "indexes more of a solidarity and co-alignment with the interlocutor, through the speaker's packaging of the referents as shared information" which is, says Strauss, an alignment strategy (ibid.: 147). Her discussion explains several features of Wilcox's speech.

As Wilcox moved through 2001, he was increasingly less able to present information and achieve transactional goals, and he was becoming increasingly dependent on exophoric and non-phoric reference. His uses of *this* in 2000 supported Strauss in its co-occurrence with new information, including when he offered an opinion about the cat who often visited his unit:

LW: [*Long pause*] There's a cat somewhere around here
LM: Is there?
BD: A cat. I saw it earlier this morning when I was here.
LM: Ah-huh
LW: I've seen it twice today.
LM: Oh, you have?
LW: Yes
LM: He's a really pretty kitty. Do you like cats?
LW: Yeah, I like them,
LM: ==oh==
LW: ==but==this ain't no place for a cat in here, he'd be out on a farm somewhere

By 2001, his uses of *this*, whether adjectival or pronominal, were all exophoric. And, while his uses of *that* were primarily exophoric or non-phoric, he was able to utilize *that*-initial formulaic phrases for relational goals such as alignment and the maintenance of positive affect between himself and his conversational partners: he sustained rapport.

Rapport, routine, and social politeness

Under normal circumstances, people engaged in conversation spend a good bit of effort on avoiding offense or its appearance, and on being and appearing competent and worthy, whether directed toward the conversation partner or one's self. On an informal or even superficial level, we could say that the speakers are engaged in saving face, or facework; in actuality, facework is complex, its components debatable, and constrained in both production and interpretation by its sociocultural context. Social politeness is, according to Rhys and Schmidt-Renfree, not keyed to the facework that supports addressing the informational needs of the conversation partner. Instead, they draw on Janney and Arndt's original distinction between tact and social politeness, where tact is creative and applicable to facework, and social politeness is conventional in nature, consisting of conversation routines learned early in life, rehearsed often, and practiced for years (535–6).

Rhys and Schmidt-Renfree claim, in their case study of "Gertie," that "Alzheimer's patients will continue to engage in social politeness after they are no longer able to be tactful" (536). Our case study of Wilcox supports their claim, as well as their suggestion that some speakers, such as Gertie and Larry Wilcox, are interested in protecting their own face. However, we suggest a slightly different emphasis, looking instead to rapport as an interactional process keyed to attaining relational goals while presenting oneself as socially competent. Rapport is seen here as the establishment and maintenance of a positive affect between speakers who feel they have an affinity by being aligned in some way, or by having some degree of unity in the form of shared knowledge or common ground, although they may not be able either to explain the basis for this feeling or point to a specific transaction of information keyed to or enabled by the rapport.

Rapport is built on small talk, formulaic expressions, conventionalized routines whereby participants in a conversation reconstruct themselves as part of the same discourse community, by mutually profferring and recognizing common discourse habits and expressions, including gestures, facial expressions, laughter. Like a grooming ritual, the mutual construction of rapport presents the illusion (and it may well be the reality) that each participant cares about, and may even guard, the wellbeing of the other. It must be co-constructed, although its creation does not have to be evenly divided; its participants may enjoy it for the period of a single conversation or for a lifetime. The creation of rapport argues a certain social competence as being able to initiated and sustain positive relational goals. Again, as the ability to handle the transactional, the propositional and the informational diminishes, we find that the emphasis on suggesting competence and sustaining the relational increases.

Wilcox furthers the construction of rapport by remaining upbeat, projecting a positive attitude about himself by interjecting laughter and humor into conversations, and by mitigating negation. The state of his health, itself a routinized, conventionalized topic coming just on the heels of a greeting routine, is a topic that generates fragments of routinized, formulaic, frozen expressions (Davis *et al.* 2000) reflecting his determination to maintain positive affect. In one of our earliest conversations, January 26, 2000, he uses a number of different fixed responses:

(9) BD: How have you been – feeling okay?
LW: Yeah. I'm improving right along.
BD: That's great.
LW: Sure is.
BD: Of the people I come here to see, a lot of them have colds and you don't – you look well.
LW: It's my iron will.

Wilcox's responses in (9) seem to reflect both a positive state of health and a positive attitude toward himself and the conversation. Moments later, however, when he is introduced to another interviewer, his remarks seem to contradict his earlier ones. In (10a), he admits to having a cold, as mentioned in (1a) above, and indicates in (10b) that his day has not been so good:

(10a) LM: My name is Linda. Do you have a cold today?
LW: I sure do – that thing's running a mile a minute.
(10b) LM: How do you feel today?
LW: Well, it hasn't been much of a day to it for me (*Laughs* [*Pause*] I'll look forward to pick up after a while.....

If measured solely in terms of transactional goals, Wilcox's responses are contradictory. In terms of relational goals, however, they are consistent. We see a similar consistency in terms of relational goals in ways Wilcox strives to maintain rapport with his interlocutors, sometimes through playful or mildly flirtatious remarks, as in (11a–d):

(11a) [1-26-00]
BD: How's your roommate, Robbie?
LW: Oh he's all right.... anytime with the ladies he's fine (*Laughter*)
(11b) [11-15-00a]
BD: I hadn't seen you today so I wanted to come by and say hello.
LW: Miss me, huh?
(11c) 11-12-01
BD: Yes, she's wonderful!
LW: I've always enjoyed the clothes she wears. I've asked her permission to whistle at her once in a while, but...
(11d) 11-15-01
LM: Oh, uh huh. You listen a lot, that's why.
LW: Honey, I ain't missing nothing!

Here, Wilcox's conversational stance is reminiscent of Sabat's (2001) example of Mrs. D's persona as "life of the party"; we find that, like Mrs. D., Wilcox uses humor as one way of maintaining self-esteem.

Humor is among 12 "indicators of relative well-being" enumerated by Kitwood and Bredin 1992 (qtd. in Sabat 2001: 109). Sabat (2001) points out what should be obvious about these indicators: "they become apparent only in social interaction between people" (p. 109). When provided opportunities for such interaction, Wilcox readily supplies the indicators. Excerpts (12a–f) run through both 2000 and 2001, as shown by the dates, and reflect Wilcox's helpfulness (12a), warmth (12b), dry wit (12c), social sensitivity (12d), and sense of relaxation (12e).

(12a) 1-12-00

BD: Yes we did. The whole tape got filled up and we turned it over, just, well, you fixed the tape and that made it work so, and we filled the whole thing up with our conversation.

LW: (*Laughs*) Well, maybe they'll give me a job fixing them then (*Laughter*)

(12b) 11-29-00

GN: Well! Have a seat and join us! Watch out for your foot!
You sit down and have a seat.

LW: When somebody wants me, it's hard to turn down!

(12c) 2-13-01

GN: Have you ever played golf?

LW: Yes.

GN: You have?

LW: Played *at* it.

(12d) 3-14-01

GN: That's right. Sure, ok, I'll come back and talk to you a little later, ok?

LW: Suits me, buddy!

(12e) 11-7-01 [*GN is showing face cards from a deck of playing cards*]

GN: Do you remember this one, Les?

LW: Yes! If there was any way to cheat, I would!

(12f) 11-7-01

BD: Is breakfast time all right? About this time of day?
Or would you rather meet in the afternoon?
Your choice:.... [*Long pause while Wilcox looks up*] how about late morning?

LW: Well, don't let me interfere with you in any way, shape or form!

As Sabat (2001) points out, it is critical that Alzheimer's speakers have the opportunity to express these indicators of relative well-being and that they be reinforced by conversation partners. Lack of healthy feedback results in "undermining a person's sense of confidence, hope, personal worth, and agency" – the four "global sentient states" identified by Kitwood & Bredin (1992).

Conversational routines reinforce connections with conversational partners. Wilcox is able to complete the kinds of formulaic routines that Rhys & Schmidt-Renfree (2000) associate with social politeness. Through such routinized speech acts exemplified by (13a–c), Wilcox affirms the presence of others and his connection with them.

(13a) *Routines of meeting, greeting, departing*
hello there, hey hey, hi, good morning, I'm fine, I'm glad to meet you, bye, goodbye to you, nice to see you too, me too, see you later, enjoyed talking to you, don't make it so long, you stay well and everything

(13b) *Routines of thanking*
you're welcome, thank you (Ma'am), I appreciate that water, you too, no thanks

(13c) *Routines of apologizing (and requesting repair)*
pardon?, beg your pardon, sorry, I don't hear well

Wilcox is sensitive to these routines in his responses to questions as well. He responds affirmatively using a range of tokens, sometimes with assurance as in (14a), sometimes with mitigated affirmatives, as in (14b):

(14a) yes, yeah, sure, I sure do, sure did, sure is, it sure is, there sure is, sure has, right, all right, it's all right with me, that's all right, right much so, fine with me, okay, absolutely

(14b) I guess, I guess so, I guess it is, I think, Must be, I reckon that's so

In 2000, and less frequently manifested in 2001, Wilcox displays several ways to indicate direct negation or that he does not know an answer (14c), and can often manipulate the intensity of his comment or initiate mitigation by expanding it, as represented by (14d)

(14c) no, nah, naw, never did, I ain't going to do that, not much

(14d) (*2000*) I don't know where he's at, I don't know – I'm scared to answer that, oh Lordy no I don't know for sure, I don't even remember that, I couldn't tell you...now my wife if she's here...she could tell you, Well now that's a good question – I don't know, I don't know that one, I don't know what to do! I don't know, confound it!
(*2001*) I don't know! I'm just wondering!

At other times, Wilcox may further mitigate his negative response, using any of several different techniques. In (14e), he uses a frozen, regional expression; in (14f), a year later, he is occasionally able to expand his negative response in a way that maximizes his social competence by signaling his awareness of the topic to his conversation partner and by continuing the conversation.

(14e) BD: Last week you were telling me that they were painting all the time to make it look really good
LW: Yes . . I told you the truth the way it was told to me
(14f) 7-3-01
BD: Hmm.
[*Long pause with no response from Wilcox*]
Been on any vacations lately? Any trip?
[*Long pause with no response from Wilcox*]
Have you been on a vacation lately?
LW: I don't know if you'd call it a vacation or not. I've just been to see my mother.
11-7-01 [*GN is showing face cards from a deck of playing cards*]
GN: The king, yes! Do you remember the king, too?
LW: Well, probably not. I can count them on my fingers or something.

When Wilcox has a problem processing a question or when he wants to avoid a direct response, he often uses modal phrases; the case studies of metadiscourse in dementia presented by Dong, Tardiff, & Ska (2003) suggest that modalizations could signal an effort to maintain communication when the speaker has narrative impairments. In (15a) Wilcox indicates that he cannot spell WORLD backwards. The modal sentence is his face-saving explanation that he could figure it out. In (15b), his use of modals adds to his mitigation of negation and is part of his avoidance of most of the topics introduced by GN.

(15a) LW: No I hadn't thought about it
LM: okay. and
LW: I guess I could sit down word for word and figure them out
(15b) GN: How's life these days, [Larry]?
I've said good morning to everybody, and now I'm talking to you.
Do you remember the tape recorder?
[*Long pause with no response from Wilcox*]
It's running, isn't it?

It's turning in here. [*Points to tape in the recorder*]
Did you enjoy your breakfast?
LW: I [would] rather see us get washed and shaved.
GN: Oh!
LW: I'll see you in Pennsylvania.
GN: In Pennsylvania?
LW: In the meantime, I'll be getting packed.
GN: Why Pennsylvania? Is that a good place to be at?
LW: I think it would. You gotta be at, of course, it would be better than to go out.
GN: Yes. Any special place in Pennsylvania?
Out in the country or in the city?
Philadelphia or Pittsburgh?
LW: Yes, I believe you'd be better off.
GN: It may be chilly though, at this time of year!
It's better up north?
LW: You can tell that. Have the signal ready. Twist them bolts if you want to.
GN: Yes. Did you drive to Pennsylvania?
LW: No, I didn't. But he did.
GN: I see. How many of you went?
Was there a group or just the two of you?
LW: I'll take a little more off of the play. Joe Lewis is right there!
GN: I see. (*Laughs*) Any place that you went fishing out there?
Did you go out and work there?
In looking at this picture, does this remind you of anything back in Pennsylvania?
[*Long pause with no response from Wilcox*]
You are having problems getting to what you are referring to.
[*Long pause with no response from Wilcox*]
Can you help me to see the context of the situation you are in?
[*Long pause with no response from Wilcox*]
Hmmmm....
[*Long pause with no response from Wilcox*]
So does this picture remind you of anything in Pennsylvania?
[*Long pause with no response from Wilcox*]
Ok. I think I'll come back in a few minutes, ok?
LW: Can I possibly get a drink?
GN: Oh, yes! That would be ok!

Closing comments

Our case study of Larry Wilcox has examined how he combines referentiality and politeness for indirect communication of information and direct efforts to achieve relational goals through conversational interaction as he moves from moderate to more severe cognitive impairment. We have noted the importance of analyzing conversation in natural settings as complementary to clinical testing, and have focused initially on looking for where and how Wilcox signals awareness of social as well as informational limitations, in order to understand ways he and his conversation partners create referentiality in interaction. We have interrogated the definition of "empty words," finding that words such as "*thing*" or "*anything*" often have social-relational functions, and we have reviewed the functions of formulaic phrases with indefinite expressions, identifying baseline functions with non-impaired older adults from the same geographic region.

We have then analyzed how certain markers of cohesion, namely demonstrative pronouns, can be used to sustain social interaction even as their speaker grows dependent on exophoric and non-phoric reference. Finally, in agreement with Ripich *et al.* (2000) on how speakers like Wilcox move to compensate pragmatically for informational or cognitive loss, we examine the initiation and maintenance of rapport as an interactional strategy that compensates for missing skills and referents.

In a study of the collaborative construction of joking by native and second-language speakers, Davies says that rapport is a specialized type of involvement as defined by, e.g., Gumperz (1982a and b) Chafe (1982) and Tannen (1989). Her claim is that joking supports indirect communication between first and second-language speakers in that it "reveals the fine-tuning of understanding" by "playing within the frameset of the other" and is thereby important in establishing rapport. Like the second-language learners delineated by Davies, speakers whose Alzheimer's disease has resulted in moderate to more severe forms of cognitive decline are limited in expression, and they, too, must deploy any lexical, pragmatic and social-interactional skills that they can access in order to achieve a momentary alignment in support of what we would emphasize are sometimes transactional but are always relational goals.

An increasingly fine-tuned analysis of the conversational skills needed to achieve relational goals is, we think, important. First, interventions incorporating such an analysis can help caregivers identify cues mentioned by Dijkstra *et al.* that can be used to scaffold the construction of an interaction, given that conversation, as they note, may be the only real contact an Alzheimer's speaker, especially for those in nursing homes, might have with others:

> When talking with persons in later stages of dementia, simply repeating information, encouraging the person with dementia to keep talking, providing cues, and using external memory aids (Bourgeois *et al.* 2001; Dijkstra *et al.* 2002) may help improve their conversational skills substantially (2004: 276–7).

We can more effectively collaborate in rapport-building as well, by assuming that regardless of the inability to sustain informational exchange, the Alzheimer's speaker remains concerned to appear socially competent. Bou-Franch & Garces-Conejos (2003) quote Thomas on pragmatic failure as it impinges on the second-language learner:

> While grammatical error may reveal a speaker to be a less than proficient language-user, pragmatic failure reflects badly on him/her as a *person*. (Thomas 1983: 97, qtd Bou-Franch & Garces-Conejos 2002: 2).

At early and moderate stages of the disease, many Alzheimer's speakers express awareness of their increasing problems with communication and memory and, as the disease progresses, some continue to use parts of their pragmatic repertoire to signal both some degree of awareness and some degree of regret. Caregivers can help Alzheimer's speakers to see themselves as persons with some competence by initiating or recognizing and sustaining efforts at rapport.

Appendix

Fall 2001. LW is sitting on a two-cushion sofa. GN is bending over, trying to make eye contact. GN holds a calendar with pictures of North Carolina scenic areas for the different seasons.

GN: No, I see. Is there anything special that you would like to do?
Like driving somewhere out in the country?
Would that be good?
[*Extended pause, with no response from LW*]
Have you ever been to Grandfather Mountain?
Do you remember ever going there?
It has a bridge at the top of the mountain.
Have you been there?

LW: [*Nod*]
GN: You have? Did you like it?
LW: I guess I did.
GN: (*Laughs*) Did you do some trekking, too?
Or did you just drive up and enjoy the view?
[*Extended pause, with no response from LW*]
Did you know it's starting to snow up there now?
LW: No, I didn't know.
GN: You didn't? I saw a picture of it. It's so cold up there, that they have seen the first snow.
It's getting colder now.
Do you like snow?
LW: I spent the weekend up here.
GN: Oh, did you?
LW: Last year. They told me, "You can't lose that!" I said, "A hurricane!" [*Non-phoric*]
GN: A hurricane?!
LW: I'm in it!
GN: Oh!
LW: Was that the creek? [*Exophoric, as LW looks at the calendar GN has brought*]
GN: Possibly, yes.
It may have been a big problem if there's a hurricane.
Lately there hasn't been any.
Do you remember Hurricane Hugo?
LW: Come slide up and sit down so you don't have to strain yourself like that.
GN: You think I could come a little closer and sit more like this?
LW: (*Laughs*)
GN: Do you remember Hurricane Hugo?
LW: A real tall fella? I believe I do

References

Almor, A. (1999) "Noun-phrase anaphora and focus: the Informational Load Hypothesis". *Psychological Review* 106: 748–65.

Almor, A., Kempler, D., MacDonald, M., Andersen, E., & Tyler, L. (1999) "Why do Alzheimer patients have difficulty with pronouns? Working memory, semantics, and reference in comprehension and production in Alzheimer's disease." *Brain and Language*, 67: 202–27.

Biber, D., Johansson, S., Leech, G., Conrad, S., & Finegan, E. (1999) *The Longman Grammar of Spoken and Written English*. London: Longman.

Bou-Franch, P. & Garces-Conejos, P. (2003) "Teaching linguistic politeness: A methodological proposal." *IRAL* 41: 1–22.

Bourgeois, M.S., Dijkstra, K., Burgio, L., & Allen-Burge, R. (2001) "Memory aids as an AAC strategy for nursing home residents with dementia". *Journal of Augmentative and Alternative Communication* 17, 196–210.

Brown, P. & Levinson, S. (1987) *Politeness: Some Universals in Language Usage.* Cambridge: Cambridge University Press.

Chafe, W. (1982) "Integration and Involvement in Speaking, Writing, and Oral Literature". In Tannen, D. (ed.), *Spoken and Written Language: Exploring Orality and Literacy.* Norwood, NJ: Ablex, 35–53.

Cherney, L., Shadden, B., & Coelho, C., eds. (1998) *Analyzing Discourse in Communicatively Impaired Adults.* Gaithersburg, MA: Aspen.

Coelho, C.A. (1998) "Analysis of Conversation". In Cherney, L., Shadden, B., & Coelho, C. (eds.) *Analyzing Discourse in Communicatively Impaired Adults.* Gaithersburg, MA: Aspen.

Coelho, C., Youse, K., Le, K., & Feinn, R. (2003) "Narrative and conversational discourse of adults with head injuries and non-brain-injured adults: a discriminant analysis." *Aphasiology* 17, 499–510.

Davis, G. & Coelho, C. (2004) "Referential cohesion and logical coherence of narration after closed head injury". *Brain and Language* 89: 508–23.

Davies, C. (2003) "How English-learners joke with native speakers: an interactional sociolinguistic perspective on humor as collaborative discourse across cultures." *Journal of Pragmatics* 35: 1361–85.

Davis, B., Moore, L., & Peacock, J. (2000) "Frozen phrases as requests for topic management: effect and affect in recipient design by a speaker of Alzheimer's discourse." Presented at *NWAV2000*, Michigan State University.

Dijkstra, K., Bourgeois, M., Allen, R., & Burgio, L. (2004) "Conversational coherence: discourse analysis of older adults with and without dementia." *Journal of Neurolinguistics* 17: 263–83.

Dijkstra, K., Bourgeois, M., Petrie, G., Burgio, L., & Allen-Burge, R. (2002) "My recaller is on vacation: discourse analysis of nursing-home residents with dementia" *Discourse Processes* 33: 53–76.

Duong, A., Tardif, A., & Ska. B. (2003) "Discourse about discourse: What is it and how does it progress in Alzheimer's disease?" *Brain and Cognition* 53 (2003) 177–80.

Durkheim, E. (1915) *The Elementary Forms of the Religious Life.* NY: Free Press.

Fox, C. & Thompson, S. (1990) "A discourse explanation of the grammar of relative clauses in English conversation." *Language* 66: 51–64.

Fraser, B. (1990) "Perspectives on politeness." *Journal of Pragmatics* 14: 219–36.

Goffman, E. (1981) "Footing." In Goffman, E. (ed.), *Forms of Talk.* Philadelphia: University of Pennsylvania Press, 124–57.

Goodwin, C. (1995) "Co-constructing meaning in conversations with an aphasic man." *Research on Language in Social Interaction* 28: 233–60.

Goodwin, C. (ed.) (2003) *Conversation and Brain Damage.* Oxford: Oxford University Press.

Grainger, K. (2002) "Politeness or impoliteness? verbal play on the hospital ward. *Sheffield-Hallam Working Papers: Linguistic Politeness and Context.* Online at http://www.shu.ac.uk/wpw/politeness/.

Guendouzi, J. & Mueller, N. (2001) "Intelligibility and rehearsed sequences in conversations with a DAT patient." *Clinical Linguistics and Phonetics* 15: 910–15.

Gumperz, J., 1982a *Discourse Strategies.* Cambridge: Cambridge University Press.

Gumperz, J., 1982b *Language and Social Identity*. Cambridge: Cambridge University Press.

Halliday, M. (1978) *Language as Social Semiotic*. London: Edward Arnold.

Halliday, M. & Hasan, R. (1976) *Cohesion in English*. London: Longman.

Hamilton, H. (1994a) *Conversations with an Alzheimer's Patient*. Oxford: Oxford University Press.

Hamilton, H. (1994b) "Requests for clarification as evidence of pragmatic comprehension difficulty: the case of Alzheimer's Disease". In Bloom, R., Obler, L., De Santi, S., & Erlich, J., eds., *Discourse Analysis and Applications: Studies in Adult Clinical Populations*. Hillsdale, NJ: Lawrence Erlbaum, 185–99.

Hengst, J. (2003) "Collaborative referencing between individuals with aphasia and routine communication partners." *Journal of Speech, Language, and Hearing Research* 46: 831–49.

Hier, D., Hagenlocker, D., & Schindler, A. (1985) "Language disintegration in dementia: effects of etiology and severity." *Brain and Language* 25, 117–33.

Hymes, D. (1972) "Models of the Interaction of Language and Social Life". In Gumperz, J. & Hymes, D. (eds.) *Directions in Sociolinguistics*. NY: Holt, Rinehart & Winston, 35–71.

Kempler, D. (1991) "Language Changes in Dementia of the Alzheimer Type". In Lubinski, R., Orange, J., Henderson, D., & Stecker, N. (eds.), *Dementia and Communication*. Philadelphia, Hamilton: B.C. Decker, 98–114.

Kintsch, W. & Van Dijk, T. (1978) "Toward a model of text comprehension and production." *Psychological Review* 85, 363–94.

Kitwood T. & Bredin K. (1992) "Towards a theory of dementia care: personhood and well-being". *Ageing and Society* 12, 269–87.

Liles, B. & Coelho, C. (1998) "Cohesion Analysis". In Cherney, L., Shadden, B., & Coelho, C. (eds.), *Analyzing Discourse in Communicatively Impaired Adults*. Gaithersburg, MA: Aspen, 65–84.

Merrison, A. (2002) "Politeness in task-oriented dialogue". *Sheffield-Hallam Working Papers: Linguistic Politeness and Context*. http://www.shu.ac.uk/wpw/politeness/.

Nicholas, M., Obler, L., Albert, M., & Helm-Estabrooks, N. (1985) "Empty speech in Alzheimer's disease and fluent aphasia". *Journal of Speech and Hearing Research* 28: 405–10.

Overstreet, M. (1999) *Whales, Candlelight, and Stuff Like That: General Extenders in English Discourse*. NY: Oxford University Press.

Overstreet, M. & Yule, G. (2000) "Formulaic disclaimers." *Journal of Pragmatics* 32: 45–60.

Perkins, L., Whitworth, A., & Lesser, R. (1998) "Conversing in dementia: a conversation analytic approach." *Journal of Neurolinguistics*, 11: 33–53.

Ramanathan-Abbott, V. (1994) "Interactive differences in Alzheimer's discourse: an examination of AD speech across two audiences." *Language in Society* 23: 31–58.

Ramanathan, V. (1995) "Narrative well-formedness in Alzheimer's disease: an interactional analysis across settings." *Journal of Pragmatics* 23: 395–419.

Ramanathan, V. (1997) *Alzheimer Discourse: Some Sociolinguistic Dimensions*. Mahwah, N.J.: Lawrence Erlbaum.

Reisberg, B., Ferris, S., Leon, J., & Crook, T. (1982) "The global deterioration scale for the assessment of primary degenerative dementia". *American Journal of Psychiatry*, 139: 1136–9.

Reisberg, B. (1988) "Functional Assessment Staging (FAST)." *Psychopharmacology Bulletin* 24: 653–9.

Rhys, C. & Schmidt-Renfree, N. (2000) "Facework, social politeness and the Alzheimer's patient." *Clinical Linguistics and Phonetics* 14: 533–43.
Ripich D., Carpenter, E., & Ziol, E. (2000) "Conversational cohesion patterns in men and women with Alzheimer's disease: a longitudinal study." *International Journal of Language and Communication Disorders* 35: 49–64.
Ripich, D., Fritsch, T., Ziol, E. & Durand, E. (2000) "Compensatory strategies in picture description across severity levels in Alzheimer's disease: a longitudinal study." *American Journal of Alzheimer's Disease*, 217–27.
Ripich, D. & Terrell, B. (1988) "Patterns of discourse cohesion and coherence in Alzheimer's disease." *Journal of Speech and Hearing Disorders* 53: 8–15.
Sabat, S. (2001) *The Experience of Alzheimer's Disease: Life through a Tangled Veil.* Blackwell: Oxford.
Sabat, S. (2002) "Surviving manifestations of selfhood in Alzheimer's disease: a case study." *Dementia: The International Journal of Social Research and Practice* 1: 25–36.
Sabat, S. & Cagigas, X. (1997) "Extralinguistic communication compensates for the loss of verbal fluency: a case study of Alzheimer's disease." *Language and Communication* 17: 341–51.
Sabat, S. & Harré, R. (1992) "The construction and deconstruction of self in Alzheimer's disease." *Ageing and Society* 12, 443–61.
Sabat, S. & Harré, R. (1994) "The Alzheimer's disease sufferer as a semiotic subject." *Philosophy, Psychiatry, and Psychology* 1: 145–60.
Schiffrin, Deborah. (1994) *Approaches to Discourse.* Oxford: Blackwell.
Scollon R. & Scollon, S. (2002) *Intercultural Communication.* 2nd edn. Oxford: Blackwell.
Shadden, B. (1998) "Sentential/Surface-Level Analyses." In Cherney, L., Shadden, B., & Coelho, C. (eds.) *Analyzing Discourse in Communicatively Impaired Adults.* Gaithersburg, MA: Aspen.
Shakespeare, P. (1998) *Aspects of Confused Speech: A Study of Verbal Interaction between Confused and Normal Speakers.* Mahwah, N.J.: Erlbaum.
Solomon P., Hirschoff, A., Kelly B., Relin, M., Brush, M., DeVeaux, R., & Pendlebury W. (1998) "A 7 minute neurocognitive screening battery highly sensitive to Alzheimer's disease". *Archiv Neurologia* 55: 349–55.
Strauss, S. (2002) "*This, that,* and *it* in spoken American English: a demonstrative system of gradient focus". *Language Sciences* 24: 131–52.
Tannen, D. (1989) *Talking Voices: Repetition, Dialogue, and Imagery in Conversational Discourse.* Cambridge: Cambridge University Press.
Tannen, C. (1999) "New York Jewish conversational style." Excerpted and reprinted from (1981) *International Journal of Sociology of Language* 30: 133–49, with pagination to A. Jawaorski and N. Coupland, eds. *The Discourse Reader.* London and NY: Routledge, 459–73.
Temple, V., Sabat, S., & Koger, R. (1999) "Intact use of politeness strategies in the discourse of Alzheimer's disease sufferers." *Language and Communication* 19: 163–80.
Ulatowska, H. & Chapman, S. (1991) "Discourse Studies". In Lubinski, R., Orange, J., Henderson, D., & Stecker, N. (eds.) *Dementia and Communication.* Philadelphia, Hamilton: B.C. Decker, 115–32.
Wray, A. & Perkins, M. (2000) "The functions of formulaic language: an integrated model". *Language & Communication* 20 (1): 1–28.
Wray, A. (2002) *Formulaic Language and the Lexicon.* Cambridge: Cambridge University Press.

5

Carousel Conversation: Aspects of Family Roles and Topic Shift in Alzheimer's Talk

Jeutonne P. Brewer

Introduction

Conversations with an individual with Alzheimer's disease, or dementia of the Alzheimer's type (DAT), particularly a close relative, are sad because they are disconcerting, difficult, and often disorienting for the family caregiver. Caregiver and DAT relative often use different frames and different discourse rules. There can be humorous exchanges because the speakers' frames don't match and they expect different results, that is, they no longer share "family talk" in the same way. The DAT individual operates in a discourse sphere that is often disconnected from earlier family talk experiences. The son or daughter or spouse must try to locate that sphere, try to function in unclear frames, and learn to follow different, and even strange, rules in order to cooperate in the communicative process. It is like trying to sing a favorite song in a minor key.

The caregiver is often disconcerted and frustrated by the frequent repetitions of words, phrases, and frames in conversations with a DAT relative. An image I use to explain such discourse is a merry-go-round – a carousel conversation. Caregivers feel like they are going in a circle. The same subject comes up. They go over and over the same or similar ground again and again. Then the conversation may end or it may pause. It's like having a pause button on the carousel. When someone presses the pause button, the particular conversation will stop, and it will start again at some point. The caregivers get another ride. Caregivers often think of such conversation as a strange linguistic ride.

I write this chapter with a special sense of poignancy. During the past 15 years, I have lived through five dementias. My husband, Chris, and I served as the caregivers for both his parents and my parents as they lived with the effects of dementia. My father, Bill, suffered from arteriosclerosis which led to serious problems when he focused his demented anger on the person he most loved, his wife. My mother, Ila, struggled with the effects of Transient Ischemic Attacks and Parkinson's Disease; this energetic and vibrant woman eventually lost her mobility and consciousness to a dementia of the Alzheimer's type. My father-in-law, Clarence, experienced weakness, instability, and mental confusion before he had a stroke; within two months kidney failure brought on an agonizing but mercifully brief end-of-life suffering. My mother-in-law, Ruth, was diagnosed with Alzheimer's disease in 1984; as with other adversity in her life, she fought hard against the effects of this disease, but it was a battle she couldn't win. Now my husband, Chris has been diagnosed with Alzheimer's disease. Like his mother, he struggles with and against the effects of the disease. However, three retinal surgeries recently took much of his eyesight, and Alzheimer's has taken much of his mental ability to plan, organize, and work on do-it-yourself projects. He can still walk around the home and property he loves, at least for short periods of time.

I am not writing a research chapter, although I do draw upon research methodology and my experience as a teacher and researcher in linguistics. Primarily, I will present my real life experience as a participant-observer with the five dementias that have been part of my life. I think that all these experiences are a carousel through reality.

Background and context

My chapter will focus on conversations between Ruth, an 86-year-old woman diagnosed with dementia of the Alzheimer's type (DAT), and her only son, Chris. The context is difficult and strained because Chris must tell his mother that her husband of 68 years has died. The linguistically complex exchange is simultaneously sad, humorous, and surreal.

Perhaps we can better understand the disconnect and discordance in their efforts to converse if we better understand the DAT individual before the disease interferes with family talk. Ruth's age and the death of her spouse are only a small, although important, part of her story. The eighth of 13 children, Ruth worked on the family farm and picked cotton. She taught school when her older brother, the regular teacher, was ill. Ruth told me one time that she learned that she didn't want to be a teacher; she just didn't have the patience. In summary, she grew up in Arkansas, married a young man who enjoyed working with that new technology, the automobile, and bore three children, two girls and one boy. During part of the Great Depression years the family lived in the small town of Grubbs, where Ruth helped feed the family by raising chickens, selling some milk from her one-cow herd, and farming the land not occupied by the house on the two lots in Grubbs. During most of her adult life, Ruth worked as a breakfast cook or waitress. She never earned more than $35.00 a week plus tips, a typical salary for women in small-town restaurants. In those days, there was no minimum wage for workers in her situation. She chose to work the early shift. She arrived at 5:00 a.m. and opened the café or restaurant at 6:00 a.m. By working this shift, she was able to be at home when her youngest child got home from school. Ruth was energetic, hard-working, and frugal. She contributed to her family's income, always saved part of what she earned, and managed to send her youngest child to "the university." Ruth was a pillar of her family – not the decorative type on the front of the building but the stable type that supports the building.

Ruth lived in Arkansas until she retired. In 1968 she and her husband, Clarence, moved to North Carolina to live with their son. But people like Ruth don't retire; they just move on to other activities. She became the family gardener – and farmer. Her son had a new yard which she groomed to perfection. The family always had an abundance of squash, green beans, tomatoes, and strawberries – sometimes cantaloupe and potatoes. She painted the yard with roses, tulips, and dahlias. She proved that one old saying could be true: she could plant a dead stick and make it grow. Mother and son always enjoyed working in the yard together. Although Ruth was supposed to let her son mow the yard, she usually mowed the grass during the day so that the job would be done exactly as she wanted. Then she would get her son to help her plant trees or a new flower bed or line the walk with plants. Only a few times did she work with soil turned by a tiller. She preferred her shovel. During a period of about 20 years living with us, she wore out three shovels. Not the handles. We didn't count those. She always wore the shovel blade paper thin before she told her son that she needed a new shovel.

Things began to change in the mid-1980s. Sometimes she said she felt tired, forgot things, and felt "goofy" after working in the yard. In 1984 she suffered a serious bout with the flu. As she was recovering, she stood by the kitchen window one morning in March and asked why the trees were all red. We called the doctor, who adjusted her medication. The green trees returned, but Ruth was already traveling down the DAT road. Specialists at the Duke Center confirmed her family doctor's diagnosis. Ruth continued to live with her family for five more years, but she gradually lost the ability to work in the yard and her interest in gardening. Her body remained strong, but she gradually lost her mobility and her ability to reason and think. At the end of her life she was taking one medication – an aspirin a day for her arthritis. For almost 12 years she struggled with Alzheimer's before losing the battle in 1995 at age 93.

Ruth's problems at home included trying to remember her regular routine, to recall names, to remember her relationship to the people with those names. At times she accused her husband, Clarence, of taking or hiding her purse. One time she hid her purse so successfully in the carport that the family had to search all day; at that point Chris removed her identity cards, medical cards, and all but a few dollars from the purse. Although she forgot many things, Ruth never forgot the name of our dog, Boots, her close companion in the yard for many years.

We meet Ruth here at the midpoint in her Alzheimer's journey. She wandered around the house but showed little interest in wandering away from home. She needed help with dressing and bathing. We had tried a number of frequently-recommended techniques for gently helping her through a day's activities. For example, we put notes on cabinets and closets to help her find what she wanted. We wrote names on the tea glasses so that she could find her place for meals. Nothing worked for us. We learned much, but Ruth was oblivious to the signs, the names, the paths we made through the house.

The year 1989 was a turning point in Ruth's life. The day after their 68th wedding anniversary in May, her husband, Clarence, suffered a stroke. From the hospital he was transferred to a nursing home ("long-term rehabilitation center"). In June Ruth celebrated her 86th birthday. A month later her husband died.

This is the context of the recorded family conversations included in this chapter. These three days of conversation are part of a collection of more than 70 hours of family conversations recorded over a five-year period.

The talking plays on

In carousel conversation the old song, "The Band Plays On," could be more appropriately titled "The Talking Plays On," because, like carousel music the talk just keeps going and going and going in circles as the music plays on. That is, the conversation starts and it keeps going until at some point that the caregiver doesn't know the DAT relative pushes the off button; that's the end of it. At least for a period of time. That may be two minutes or two hours or two days. Almost certainly the talk will start playing again and go through the routine again. The speakers may or may not go back to where they were in the earlier conversations, but they will go back to the area where they were. Of course, talk will never be exactly the same, but it will be so close as to be viewed as being essentially the same, at least by the caregiver. For the person with Alzheimer's, it's as though the conversation never took place. The DAT relative will view it as a new topic, a new conversation, a new song. So it's new – new to the DAT relative and a carousel ride for the caregiver.

The following exchange between Ruth and Chris, her son, took place on July 7, 1989, the day after Ruth's husband, Clarence, had died. Ruth and Chris had a very similar effort to converse the previous evening. Here, Chris is trying again to explain to Ruth that Clarence has died.

[RB = Ruth; CB = Chris, Ruth's son; JB = Jeutonne, Chris's wife]
(July 7: Part 1)
CB: Momma
RB: Huh
CB: Do you remember who Clarence Brewer was?
RB: Clarence Brewer? He was my husband I reckon
CB: He was? He was my Daddy too, right?
RB: That might be possible
CB: Do you remember me telling you yesterday that he died?
RB: Somebody died.
CB: It was Pop. My Daddy. Your husband. You remember me tellin you that?
RB: I heard about it somewhere.
JB: What'd you hear? What did you hear?
RB: I don't remember
CB: Do you understand what I'm telling you?
RB: I think I was.. saw something or other, but I don't remember what it was. I might 'a dreamed it.

Chris presses the pause button for the carousel conversation. He and Jeutonne, his wife, try to help Ruth remember Clarence by talking with her about his room and the old car that he drove. This doesn't help. It was important to Chris to feel that he had done everything possible to help his mother understand what was going on. He did not want her to be startled or apprehensive if one of her flashes of lucidity occurred during the funeral. So, Chris returns to the subject of his dad. The carousel starts again, and the music changes.

While I have identified and numbered the key segments of the topical thread for Clarence's death, I will not insert commentary between sections of the sequence, as is typically offered in conversation and discourse analysis. I hope that readers can thereby vicariously experience the vacillating motions of the carousel of conversation.

(*July 7: Part 2*)
CB: We're going to have a funeral for him Monday.
RB: Monday
CB: Yeah
RB: That's next week isn't it?
CB: Yeah, unhuh. Today's Friday then there's Saturday and Sunday and we're going to have his funeral Monday.
RB: Where at?
CB: Over in Greensboro
RB: What's the matter nobody hadn't told me?
CB: I told you yesterday
RB: I didn't hear you
CB: You didn't
RB: No
CB: Okay

(*July 7: Part 3*)
RB: Well can't be nothin' helped about it
CB: That's true. I just wanted to be sure you knew about it. That's why I'm telling you again. Okay?
RB: Well who's to blame for it?
CB: Nobody's to blame for it
RB: Unh
CB: There is no blame

(*July 7: Part 4*)
RB: He's not dead yet, is he?
CB: Yes, he's dead. He died yesterday
RB: He did?
CB: Uh huh
RB: Y'all didn't tell me last night
CB: Yes I did
RB: No . . . I didn't know anything about it
CB: I told ya last night, you just forgot
RB: I must a been asleep
CB: No, you was awake

RB: No ... I don't forget stuff like that
CB: You don't? Do you know who he was?
RB: Um

(July 7: Part 5)
CB: Did you ever know him?
RB: Did I ever know him?
CB: Yeah
RB: Who?
CB: Clarence Brewer
RB: Well wasn't that his name?
CB: Yeah, but who was he?
RB: Who was he ... Clarence Brewer!
CB: Yeah, but I mean, what was he to you?
RB: I don't know whether he was yours or whose
CB: You don't know if he's what?
RB: He might have been your boy
CB: No, he wasn't my boy
RB: Why?
CB: Well he was a little bit older than I am, he couldn't have been my boy. Do you remember him being your husband?
RB: I don't remember anything about that
CB: You don't. Did you know he was my Daddy?
RB: No, I don't know him
CB: You didn't. You don't remember him being your husband. Okay [*Long pause*]
RB: Well, what are you going to do about it?
CB: We're going to have a funeral for him Monday
RB: Well, I can't help that either.
CB: And you're going to go. You're going to go to the funeral with us

The carousel stops. Ruth is tired. Ruth and Jeutonne slowly walk down the hall to Ruth's bedroom.

Changing family roles

On a carousel ride, the task of the parent or other adult is to stand by the carved wooden horse and make sure the child is safe and secure. The child holds the reins and enjoys the ride. The adult will often point out waiting family so the child can wave as the carousel revolves. In a family conversation between parent and child, the question is who functions as parent and who as child? And how do you know?

In the previous section, Chris clearly served as adult in charge of these exchanges. He had one primary goal – to help his mother understand that her husband of 68 years had died. He knew that his mother had been losing her memory of Clarence during the two months he lived in the nursing home. Although Chris and Ruth regularly visited Clarence, Ruth frequently forgot why they were going to the nursing home. Clarence was absent from her home environment; he was fading from her memory.

The evening before the July 7 exchange, Chris tried to explain to his mother that her husband and his father had died. He was not successful. On July 7, Chris tried again, as seen in the sequence of segments. Acting sometimes as child, but more often as parent or authority, he asked his mother questions to help her recall their family relationships:

> Do you remember who Clarence Brewer was?
> He was my Daddy too, right?
> It was Pop. My Daddy. Your husband. You remember me tellin you that?
> Do you know who he was?
> Yeah, but I mean, what was he to you?
> Did you know he was my Daddy?

Ruth's responses ranged from uncertainty to denial: these responses show her in a subordinate role, presenting hedges and uncertainty markers.

> Clarence Brewer? He was my husband I reckon
> Somebody died.
> He's not dead yet, is he?
> I don't know whether he was yours or whose
> He might have been your boy
> I don't remember anything about that

However, it is also important to note that Ruth frequently asserted a parental authority:

> What's the matter nobody hadn't told me? (July 7, Part 2)
> Well who's to blame for it? (July 7, Part 3)

Y'all didn't tell me last night (July 7, Part 3)
I don't know whether he was yours or whose (July 7, Part 5)
Well, what are you going to do about it? (July 7, Part 5)

Chris had to function simultaneously as parent and child. Each time it seemed that Ruth understood what had happened, she said something that showed her confusion about an event or relationship. On July 8, Ruth, Chris, Jeutonne, and Clarice, Jeutonne's sister, went to the funeral home. After viewing the body, Chris and Jeutonne went to the office to complete the funeral arrangements. Ruth and Clarice decided to wait in the lobby. Clarice asked Ruth what she thought and if Clarence looked all right to her.

Ruth replied, "Well, he looked all right, but he wouldn't talk to me." The carousel turns, and the music plays. One strange thing about the carousel conversation is that it is not just circular. The carousel can also move forward, backward, or sideways while it continues its circular motion. While this is "normal" conversation for the person with Alzheimer's, it is a strange sense of movement for the caregiver. Which direction is the conversation moving? Forward and circular? Backward and circular? To the left and circular? It's as though both time and space are irrelevant. And of course to the person with Alzheimer's, they are irrelevant.

This type of role switching is surely at the center of caregiver discomfort and confusion. What role is needed now? How can I serve as parent in the discussion while showing my love and concern as a son or a daughter?

Topic shift

On July 8 friends and members of Jeutonne's family have gathered. Ruth's younger daughter, Vivian, is scheduled to arrive later that day.

She responds to Chris's questions about family, using concrete details about raising children. This topic leads Ruth to the schoolhouse she attended and later to the railroad tracks which she and her brothers and sisters walked beside on the way to school. This topic leads to the house by the tracks, the house she lived in as a child. In the conversation, topic is jointly managed by Ruth and Chris, and Ruth is willing to join in the exchange.

On July 9, Ruth at first says that she doesn't remember Clarence. When she later does indicate that she remembers him, she refuses to recognize or talk about the fact that he died, which is the reason that the family is gathering.

Then Chris, Ruth, and Vivian begin to talk about family and living in the old hometown, Grubbs, Arkansas. They talk about the details of where they lived. Ruth participates freely in this topic, remembering details about the town. Referring to a class picture of teacher and students hanging in the den, Ruth and Chris talk about the Bone Cave school and its students. They talk about the house and its location in Grubbs; Ruth remembers the birth of children. This topic is arranged by time and space in her mind. However, Clarence isn't connected in her head to those spaces. Spaces between the segments represent family talk that does not include Ruth's initiations or responses.

CB: And then . . . the barn was gone. Somebody. . . .
RB: Where was we livin' then?
CB: . . . stall out there. Sat there in a little grove of trees.
RB: Where was that at? In Grubbs?
VB: Uh huh.
. . .
RB: Well, where does – put in the Baldwin house out there? The farm is down by Grubbs Creek. Can you remember the foot-bridge across Grubbs Creek?
CB: No Mama, I wasn't even born then. There are some pictures that were taken, somebody said, of the old Bone Cave Hollow school. Up at the old place? The old wood school building with all the kids there and teachers standing. All different sizes and one teacher.
VB: Yes, uh huh.
RB: That was all they had!
CB: Most of "em all were kids.
RB: That was Bone Cave. We went to school up there!
. . .
RB: Was that at Grubbs?
VB: Yes.
RB: Where did we live?
VB: We lived there in that old house.
RB: Where?
VB: That old house there in Grubbs. Brewer house.
RB: Roadhouse? Well that thing's still there. Who did that belong to?
VB: It belonged to you!
. . .
RB: Where was you raised, now?
VB: Well, I was raised mostly in Newport. I was born in Grubbs.
RB: You were born where?

VB: I was born in Grubbs.
CB: Born in Grubbs or out on the farm?
VB: I grew up in Grubbs and was kin to the Grubbs.
RB: Well, what house did you live in?
VB: It was an old house, it had a tin top.
CB: They lived in the house . . . they lived once I think, in the house across the highway from Miss Pennington's. There was a store and another house up by Whiskers Moore used to have a store; back across there?
VB: I think it was more in the middle.
RB: I guess it was there down the farm. You was living out there!
VB: I wouldn't want to live there now.
RB: Huh?
VB: I wouldn't want to live there.
. . .
VB: There's an old house with a tin roof . . . .
RB: Huh? There's a gin down there.
VB: I was born in the immediate town. I was born in Grubbs.
RB: Where?
VB: In Grubbs. Right in town.
RB: Oh well, you know, it's seeming like one of my children, maybe two of them..was any of mine born when you was?
VB: Evelyn was born before I was. She was close to three when I was born.
RB: Huh?
VB: She was close to three years old when I was born. Evelyn was.
RB: Well, we lived close to the corner by the depot.
VB: Well, we lived all over Grubbs.
RB: Had a little old house there beside of it. I don't know when the girls was born there. It wasn't a log house, it was a little ol' slab house. I got sick and they take me over there, I believe to the. . . . Cindy and what was his name? Dan? Dan I guess it was. Dan was old – older than me. And then we went over there, "cause the house we lived in was more like a shanty.

On July 9, more people are gathered at the house. There is a family conversation going on around Ruth although not specifically with her. Ruth asks questions as a way of joining the conversation. Everybody lets her join in, and their answers to her questions give her a space in the conversation and a chance to ground at least part of the conversation in those spaces in her mind.

However, Ruth's comments reveal distinctive patterns that are essential to understanding the nature of carousel conversation. Her statements indicate that she can provide information about places and spaces in her life, at least about physical places and spaces. Most of her questions signal role switching.

When her son, Chris, talks about the old, wooden school building in Bone Cave Hollow, Ruth says, "We went to school up there." When her daughter says that they lived in "that old house" in Grubbs, Ruth says, "Well that thing's still there." A short time later there is a reference to "an old house with a tin roof." Ruth adds the information that "there's a [cotton] gin down there." Essentially Ruth's comments fulfill a parental role as she gives her children more information about family place markers – the town Grubbs, the house in Grubbs, the gin, the farm, the footbridge over the creek, and Bone Cave School.

Most of Ruth's questions serve a different function. They ask for information that a child would ask a parent in order to better understand the family's history. There is no signal that a role shift is taking place. Ruth asks:

> "Where did we live?" ["that old house"]
> "Who did that belong to?" ["It belonged to you"]
> "Where was you raised now?" ["mostly in Newport...born in Grubbs"]
> "Well, what house did you live in?" ["an old house...had a tin top"]
> "...any of mine [children] born when you was/" ["Evelyn was born before I was"]

In one brief conversation, we encounter the family's talk about "back home" and "back when" in Grubbs while the mother and her children are switching roles. Her daughter's answers to these questions explained that the family lived in Grubbs in an old house with a tin roof that was owned by the parents; some of her children were born in Grubbs; and there was a child born before Vivian was born.

Ruth provides information about physical places and spaces in Grubbs. She asks for information about family places and spaces; she needs to clarify family relationships. It is interesting that she does not ask her primary caregiver, her son, these questions. Perhaps, she wanted to avoid questions about her husband. Although this is a conversation about family and where they lived, there is not one mention of Clarence, her husband.

Only adults participate in this carousel ride, perhaps to the tune of "The Carousel Polka": it is a busy, noisy ride, like most family conversations. Ruth participates in and initiates shifts in topics – town, house, location, farm, and school. The parent–child roles often change when Ruth asks questions about family relationships. In terms of the carousel metaphor, the music is sometimes loud, sometimes filled with static and distortion. The conversation is circular, but it moves forward. Sometimes the parent provides information about family history. Sometimes the child switches to a parental role to explain family spaces.

This is the basic nature of the relationship between caregiver and the parent or spouse with DAT. The caregiver has to answer several questions: What is my role? What information is my partner with DAT giving me (or asking me to provide)? Of course, this is the nature of conversation and discourse. You receive the information, and you must respond. The information comes as a package of words, syntax, and discourse features. There is not time to analyze the linguistic detail.

Add to this the complication that every person with DAT is an individual. There are no standard DAT individuals, because their life experiences are different. Their words, syntax, and discourse will reflect these differences. Although there are general characteristics in these linguistic categories, individuals shape them to serve their purposes in communicating.

Most studies of DAT have viewed words, syntax, and discourse as separate categories to be studied. However, speakers and listeners must deal with the whole package of information. Words are more than pieces of meaning; they are meaning tied to syntax – singular or plural, definite or indefinite, past or present – and discourse. We don't know the meaning of *bank* until we know its syntactic partners (The banks are closed today. Don't bank on that information.) We don't know the significance of the words and syntax until we know the discourse context. (You are planning a trip and must buy traveler's checks. You have heard that your boss will promote you general manager next week.)

In his journey with DAT, my husband is having more difficulty with words, syntax, and discourse. A recent conversation went like this:

C: Did you [*Pause*]
J: Did I what?
C: car? [*Pause*]
J: The car?
C: Yes [*Pause*] garage? [*Pause*] put up the car
J: Oh, did I put the car in the garage?
C: Yes
J: Yes, I did.

The meaning of his question is different from the exact words and syntax. We have had this conversation on a fairly regular basis recently. This is the meaning of his question: Did you close the garage door? I have learned the discourse meaning of an exchange like this. Because his words aren't always clear, I need to check to be sure that we are both dealing with the same question. Then I can give him an answer.

I learned many things from my experience with our parents when they were ill, but I also learned that DAT affects people differently. Their paths may be similar, but they will not follow the same path. My mother would be fairly stable on a plateau until a vascular event like a TIA would cause her to slip to another plateau. My mother-in-law, Ruth, went through the stages of Alzheimer's disease, slowly but steadily declining.

My husband and I read many articles and books about Alzheimer's disease and tried to learn how to help Ruth and care for her. The idea of making tags or labels to help her find things seemed like a good idea. I made labels for the kitchen cabinet doors. Ruth always asked where she was supposed to sit at mealtime, so we bought plastic glasses and wrote our names on the glasses. The result? Ruth ignored the labels on the cabinets. She always wanted to look inside the cabinet, and then she would ask where an item was. She came to the table at mealtime and asked where she was supposed to sit. Sometimes good ideas and plans just don't work for everyone. We have to try each of them to find out what is best. This is another aspect of the carousel ride – switching horses.

As my husband, Chris, makes his journey with Alzheimer's disease, his path is different from his mother's path. An engineer by training and a do-it-yourself guy by choice, he enjoyed designing and building things – tools, racks, and furniture for the home, an improved drainage system for the yard. He maintained and repaired our cars. He set up an efficient and detailed bookkeeping system so that we would know where our pennies and dollars went. As we cared for our parents during their illnesses, Chris cooked many of our meals and took over the tasks of grocery shopping and running family errands. His life is very different now. Alzheimer's disease robbed him early of his ability to do deductive and abstract thinking. He can't add a small column of numbers, figure out how to use the key to open a door, make himself a sandwich for lunch, or read his calendar. He rarely will answer the phone, and he can't dial a number to make a call.

Alzheimer's disease is often described as "the long goodbye." It is a good description in general terms, because it shields the caregiver; there is no need to explain the sad, personal details of the goodbye. However, in my experience the caregiver often doesn't understand that there is a new situation until the change has already occurred. The part of the do-it-yourself guy has already been lost before the caregiver can say goodbye. The loss changes the life of the caregiver as well as the individual with DAT and that changes the discourse context and the conversations. There are far fewer times when we can talk about current events or family or places we knew or learned about together. Plans for the future are focused on what we might be able to do now, not what we might want to do in the distant future of the next holiday, anniversary, birthday, or vacation time.

In our carousel conversation, Chris gives me clues with his words, his syntax, and his discourse context. I listen and try to understand his discourse space. The researcher-observer might be tempted to think of this as a fascinating journey in linguistic terms. From an objective view, this is true, but the journey is harder this time because I don't have my partner to help me. Sometimes either the researcher or the wife must respond; that choice is not difficult. The researcher must step aside, wait a turn, and hope the carousel conversation starts again some time in the future.

References

Brewer, J.P. (1994) "Affirming the Past and Confirming Humanness: Repetition in the Discourse of Elderly Adults. In B. Johnstone (ed.), *Repetition in Discourse: Interdisciplinary Perspectives* (Vol. 1). Norwood, NJ: Ablex, 230–9.

Hummert, M.L. & J.F. Nussbaum (eds.) (2001) *Aging, Communication, and Health: Linking Research and Practice for Successful Aging.* Mahwah, NJ: Lawrence Erlbaum.

Johnstone, B. (1966) *The Linguistic Individual: Self-Expression in Language and Linguistics.* NY: Oxford University Press.

Ramanathan, V. (1997) *Alzheimer Discourse: Some Sociolinguistic Dimensions.* Hillsdale, NJ: Lawrence Erlbaum.

6
Alzheimer's Speakers and Two Languages

Guenter M.J. Nold

Initial considerations

The language of people who suffer from Alzheimer's disease (AD) has been a field of research in medicine and psychology for a long time. A major focus of this research has been to describe specific language-related deficiencies that can help identify people with AD in contrast to people with only slightly different symptoms. The description of deficiencies has been the basis for research that links language-related deficiencies to stages of an on-going process of psychopathological deterioration and ultimately to specific lesions in the brains of AD patients (cf. Cummings, 1992; Cummings *et al.* 1985; Heindel *et al.* 1997; Kempler 1991).

In the interdisciplinary field of first and second language acquisition (often cited as L1 and L2), psycholinguists and neurolinguists have interest in aphasias and also in various types of dementia. This interest is, on the one hand, motivated by the aim to develop an empirical base for models of language and, on the other hand, by the expectation that findings about the structure of the brain will be relevant to an understanding of the processes involved in the language acquisition process (Newmeyer 1998b; Weinert 2000). Studies of language acquisition and of language deficiencies deal with such questions as the relationship between language and thought – this is the tradition of the Sapir–Whorf hypothesis (cf. Whorf 1956) – or between language and cognition in the tradition of Chomsky (Chomsky 1980). Here the language of children with specific language-related problems (Weinert 2000: 316, 339) or of people with aphasias or of people with various types of dementia (Blumstein 1988; Caplan 1988) becomes the object of an investigation that tries to base models of language acquisition or of linguistic models of language on empirical findings (cf. Grimm 2000). One of the central issues is how language is represented in the human brain. There has been a long-standing debate on whether language is organized in modules in the brain that are accessible to conscious processes of the individual or if – and possibly how – language development is related to the development of cognitive competencies more generally (Ellis 1994: 191–288, 348–96, 417–62). It is a related problem if and to what extent language or components of language are related to or dependent on thought processes.

Slobin's now-classic language development hypothesis claimed that concept development preceded language development in children acquiring their first language (Slobin 1979). In the meantime empirical research has revealed a far more complex pattern of how general cognitive processes impact on language and vice versa. Elman *et al.* show that domain-specific influences of the environment and learning and acquisition processes interact with one another and change over time; in this view, modular specializations are the result of development in line with a connectionist model (cf. Elman *et al.* 1996; Elman 2003: 431). By analogy, insights resulting from this psycholinguistic research can broaden the range of AD investigations. The focus on language deficiencies can then shift to include a perspective that more systematically considers context variables within the environment.

An ongoing shift to a process orientation has characterized a further line of linguistic research: Alzheimer's discourse. If this research deals with Alzheimer patients' linguistic deficits in comprehension and production, it combines a process orientation with a deficit approach. A typical example of this is the investigation of Almor *et al.* which tries to explain why AD patients have difficulty with pronouns in interaction with other speakers (Almor *et al.* 1999). Such studies make an important contribution to a greater understanding of the different facets of AD. The shift to process orientation becomes even more noticeable in studies of AD discourse that put emphasis on retention rather than loss. They represent a new paradigm of AD research (Ramanathan 1997; Hamilton 1994). Davis and Moore remind us that AD speakers have to make a great effort when they interact with other speakers as they have to use their remaining linguistic and communicative competences creatively. Among other things, it seems they use formulaic language, earlier regarded as a deficit in production, as a "shortcut in processing" in order to "compensate for difficulties with memory and retrieval." (Davis & Moore 2003, 119f: cf. Wray & Perkins 2000).

Investigations into Alzheimer's discourse typically focus attention on interactions among monolingual AD speakers. However, patients with a bilingual background are also affected by the disease, so that the question arises in what way AD has an effect on their interaction. As a result, there is now an increasing interest in bilingual patients with Alzheimer's disease. The diversity of focus is similar in type to that in studies of monolingual AD patients although the number of publications is limited. On the one hand, there are studies dealing with the semantic deficits and deteriorating verbal fluency of bilingual AD patients that can be observed. Meguro *et al.* (2003) find that AD patients with an L1-Japanese and L2-Portuguese background in Brazil demonstrated impaired naming ability in both languages whereas their oral reading ability was more differentiated. The greatest problems were reported for reading Kanji in contrast to reading Kana, while Portuguese irregular words were more difficult to recognize and pronounce than regular words. In another study the verbal fluency of elderly English-Afrikaans bilingual speakers was compared to that of bilingual AD subjects. The analysis was based on semantic verbal fluency tasks. It was found that the AD patients' linguistic behavior showed fewer cases of strategic code switching and that there was some relationship between age of L2 acquisition and verbal fluency (de Picciotto & Friedland 2001).

On the other hand, interest exists in the wider ramifications of AD speakers' bilingual behavior. In an early study, Hyltenstam and Stroud try to account for "why some patients seem to retain their bilingual abilities at late stages of regression, while others lose the capacity to correctly choose and uphold the appropriate language in very early phases of the disease" (Hyltenstam & Stroud 1993: 227). One of the tentative results of the study is "that there is an interaction between degree of dementia and premorbid level of proficiency" (ibid.: 236). At the same time the relationship between the degree of dementia and the problems the patients have in keeping their two languages apart is said to be complex. Frequency of use is taken into account as much as recency of use and other conditions (ibid.: 238f.). In another early investigation Obler *et al.* point out that there have been reports of bilingual demented subjects who do not show signs of language mixing. However, at the same time their systematic investigation reveals that among demented speakers violations of linguistic constraints (e.g. code-mixing within a word) are rare in contrast to violating pragmatic constraints in code-switching (1991).

Referring to Hyltenstam's study, Paradis points out that declarative memory is affected by the illness first. This implies that AD speakers who acquired their second language in a more conscious and controlled process of language learning are more liable to L2 language loss than those who acquired their L2 in ways more similar to L1 acquisition. Furthermore, it also explains more generally why AD patients' knowledge and use of words, their explicit lexical competence, shows signs of deterioration at an early stage and why they retain grammatical functions subserved by the implicit processes of procedural memory. Paradis also points out that "pathology may selectively affect L1 and L2 depending on which cerebral structures are affected" and he concludes that "patients with Alzheimer's Disease lose access to their L2 before their L1" (Paradis 2004).

As findings such as these have great consequences for potential caregivers, it is significant to investigate more specifically how Alzheimer disease impacts the linguistic behavior of speakers with a background in more than one language. Such research has to address a greater range of bilingual backgrounds and more aspects of AD among speakers with different language experiences. It is also essential to shift this research from a focus on language deficits to an emphasis on what bilingual AD patients can do. A corresponding change in L2 learning contexts has led to a new perspective in the field of language assessment (cf. The Council of Europe, The Common European Framework of Reference for Languages 2004). Here the emphasis is now on *can do* descriptions rather than the identification of errors or deficits as the main measure of language competencies. It is in line with a focus on what AD patients can do that the following study was developed. Its research questions and design are based on an analysis of Alzheimer's discourse data. They are described in detail and some major results are discussed.

Research questions

This study of AD speakers' discourse is part of a larger research project on AD discourse by Davis and others (cf. Davis & Moore 2003: 121). One integrated component of this project was developed in cooperation with Nold, the writer of this chapter. It deals with AD speakers who have a background in more than one language. As the major focus of this component of the project is on what AD speakers can do with two languages rather than what they cannot do, a qualitative study on the basis of data collection and an ethnographic analysis by an involved observer-researcher (Saville-Troike 1997: 126ff.) was considered to be appropriate. Obler *et al.* (1991: 138) recommend the administration of the traditional speech and language assessments. However, in this training project, the research questions could be only partly worked out at the outset. Personal data were minimally available or allowable and the use of data collection techniques such as a structured interview or a questionnaire was not considered to be acceptable practice by the site's administration. Accordingly, some of the questions became more specific after the data gathering began:

(1) How do AD speakers with at least some experience in German interact with a native speaker of German who is a foreigner, male, and who speaks English fluently with a non-American English accent?
(2) In what way does the language background of the AD speakers impact on the conversations?
(3) Do AD speakers have the ability to access vocabulary of a second language that they have been in contact with?
(4) Can AD speakers recover or newly develop elements of a second language by getting involved in second-language training phases?
(5) Do the conversations with AD speakers who had an experience with more than one language shed a new light on the question of whether the L2 of AD patients goes first?

Alzheimer's discourse data were collected in an Alzheimer's Unit at Pleasant Meadows over a period of four years, with a two-year subcomponent of conversations in which Nold, a visiting professor of applied linguistics from Germany, was involved with AD speakers. The conversations were carried on with approximately ten AD patients at Stages 5 and 6 of AD (Bayles *et al.* 1992). Over a period of eight months, Nold held these conversations at regular intervals almost every week. He returned to Germany, and after five months of absence, returned to the US, and resumed the conversations again over a period of 16 months. In these conversations it was discovered that three of the Alzheimer patients had had contact with more than one language. The conversations were transcribed and digitized. This study will focus on two of them, a German-American bilingual woman, and an American man of German descent.

The AD data: a German-English bilingual

Mrs. T., a woman at the age of about 80 and considered by the facility staff to be at stage 6 of AD (keyed to deterioration in her ability to continue autonomous activities of daily living), is a first-generation American with a German-Jewish background. Available information indicated that her mother tongue is German. Furthermore, she reveals that by the time she was 18 years old her family, which had lived in a farming environment in the South of Germany, managed to leave Nazi Germany. After arriving in the USA, German was still the language spoken in the family. It can be assumed from the way she behaves linguistically and from how she speaks more specifically, that English became her second language and eventually took over the place of German. The discourse data that were collected consist of three conversations, two held in 2001 and one in 2002. The first conversation was very short while the second turned out to be an extensive one.

The analysis of conversation will, on the one hand, direct attention to the linguistic features of her speech, especially to the status of her two languages and to her code-switching practices. On the other hand, it will also take into account the cultural implications of the meeting of the two partners in discourse. What has to be kept in mind in these recorded conversations between Mrs.T. and Nold is that Mrs. T. is now in touch with a professor from Germany who shares her first language. Although he is German, there is no objective reason for Mrs. T. to associate him with Nazi Germany in any way. And yet when leaving Germany, she left behind the country where her discourse partner lives.

As the conversation with Mrs T. indicates, she is able to respond to a straightforward introduction of her interlocutor with phrases that are similarly straightforward and direct. The directness involved in this exchange is not inappropriate, although the discourse norms of politeness in German culture would suggest a less businesslike approach when meeting someone older for the first or second time:

N: Yeah, good morning! You remember me? I was here last week.
T: No doubt.
N: And we'd . . . I am Guenter Nold, I'm from Germany, and people tell me that you can speak German.
T: Well, how about that.
N: Yeah, and I was here last week and you said you were so tired you wanted to stay in your bed and sleep. Today you are up so we can talk a little, can't we?
T: Uhuh. So, what would you like?

After this introduction there is a topic shift that first seems to puzzle Mrs. T. until she is finally ready to adapt to the code switches and the topic shift initiated by her interlocutor:

N: Well, can you tell me why you can talk German?
T: What?

N: You were brought up in Germany as a child? Or did your parents speak German?
T: I –
N: Pardon?
T: I don't know what I can say.
N: You don't know. Maybe you can say "Guten Morgen" (*Good morning*)?
T: Could be.
N: Guten Morgen, Mrs. T.? Can I say that?
T: Mmmh, I see.
N: You understand that, don't you?
T: I see, mmh.
N: Yeah. Wie geht es heute morgen? (*How are you doing this morning?*) Wie geht es? How are you this morning? Wie geht es? Wie gehts? (*How are you?*) Gut? (*Fine?*)
T: Probably.
N: Sehr schön, sehr schön. Haben Sie schon gefrühstückt? (*Great, great. Have you had breakfast yet?*)
T: Oh, yeah.
N: Oh, yeah. War das Frühstück gut? (*Was the breakfast O.K.?*)
T: **Ja**. (*Yes*)

The fact that Mrs T. is initially quite puzzled does not come as a surprise. It usually takes some time and effort to switch from one language to another, especially if a code switch is not really motivated by features in the context of the situation. The attempts of Nold to switch from English to German take some time before Mrs. T. is finally persuaded to accept the shift to the new topic and the change to German. The interaction shows that she is willing to listen to German utterances and able to process them and to respond although she is not yet at the point where she would switch to German in her production. However, her final "Ja" (*Yes*) can be regarded as a smooth code switch from English to German as she seems to be using a German pronunciation for the English "Yeah", which is similar to the German "Ja" (*Yes*).

What gradually emerges here is the use of formulaic speech (e.g. "Could be"; "Mmmh, I see"), which is a typical feature of AD discourse (Davis & Moore 2003: 123). In this extract of discourse these phrases function as connecting phrases that help Mrs.T. to participate in the conversation without losing face. Her utterance "Could be" is a reaction to an indirect request to express a greeting in German ("Maybe you can say..."). It allows her to avoid a German utterance. The connecting phrases are typically short, semantically vague so that they can be employed in many situations. This does not imply that they are to be seen negatively. On the contrary, they show that Mrs. T. can use the phrases at her disposal very efficiently from a socio-pragmatic perspective.

In line with accommodation theory in sociolinguistics Mrs. T.'s phrases appear to underline the notion that people "attempt to converge linguistically toward the speech pattern believed to be characteristic of their recipients when they... desire their social approval and the perceived costs of so acting are proportionally lower than the reward anticipated" (Spolsky 1989: 108). However, the success of using fixed phrases here also depends on the perception of the partner in discourse. Giles and Powesland remark that "if accommodation is attributed to external pressures rather than voluntary effort, then it is likely to be less effective" (Giles & Powesland 1997: 238). This means the interpretation of how an AD speaker interacts is also a reflection of how the discourse partner reacts to an AD speaker and how he or she perceives and constructs the interaction for him- or herself. This involves a special interactive competence on the part of people who take part in AD conversation.

As the conversation continues Mrs T. is more and more willing to switch to German in her language production as the following extract from the same conversation demonstrates:

N: ... Wissen Sie noch eine Stadt in Deutschland, an die Sie manchmal denken? (*Are you still aware of a city that you sometimes think about?*)
T: Uhuh.
N: Welche ist dieses? Ich bin aus Frankfurt.(*Which one is it? I am from Frankfurt.*)
T: All right.
N: Welche Stadt kennen Sie? (*Which city do you know?*)
T: Mmmh.
N: Eine Stadt in Deutschland? (*A city in Germany?*)
T: Deutschland, uhuh.
N: War das Heidelberg oder Berlin oder Hamburg? (*Was it Heidelberg or Berlin or Hamburg?*)
T: Oh.
N: You know, you remember any of these places in Germany?
T: No, we have had Augsburg (*Augsburg – articulated with German pronunciation*)
N: Oh, Augsburg?
T: and Munich.

N: Oh, I see.
T: Uhuh.
N: I have a friend living in Augsburg, . . . Augsburg hat jetzt auch eine Universität.
T: Oh, really?
N: Yeah. Wo haben Sie in Augsburg gelebt, oder? (*Where did you live in Augsburg, or?*)
T: Nein, . . . (*No . . .*)
N: You can speak English if you prefer English.
T: Es kommt nur auf von, von Augsburg Hundertzehn, now mein father war ein, ein, (*It only comes (I can only remember?) from, from Augsburg one hundred and ten, now my father was a, a, a . . .*)
ein father, ein, ein f. . . . He made everybody phone, two thousand people out, see. He was one of the . . . all the things he . . . and so he had the
N: So he was the one who emigrated, or had to go and had to leave and went to America, your father?
T: Yeah, uhuh. Ja.
N: So he was brought up, he was born in Augsburg,
T: Augsburg. (*German pronunciation*)
N: or close to Augsburg, and then he emigrated to America.
T: Yes, something like Augsburg. (*Augsburg in German pronunciation*)
. . .
N: Und Ihre Familie? (*And your family?*)
T: Well, wir sind nur in Jellingen, (*Well, we are only in Jellingen.*)
N: Ah, ja.
T: und mein father was, mein . . . mein . . . everything, everything for, for everybody. (and my father was, my . . . my)

The question arises how her way of code-switching or codemixing (cf. Baker 2001: 101) has to be judged and evaluated. It becomes obvious that Mrs. T. wants to switch to German as her statement "Es kommt nur . . ." indicates. She wants to accommodate and has problems in sticking to German. When she starts to speak German more extensively here, her use of "father" rather than German "Vater" can be interpreted as codemixing as she uses an English word right in the middle of her German utterance. Her inability to stick to German is then judged to be a violation of either linguistic or pragmatic constraints (cf. Obler 1991). However, if it is taken into account how she expresses herself in both German and English and not only in German in isolation, the conclusion is that she has linguistic problems in both languages that may have similar roots. Yet, her English functions more like a mother tongue and her German more like a second language. The following statement about her family and her father underlines this still further with regard to her use of English:

> T: I mean we had the dairy (?) in other words. And that's what he was, he was all the things, that's all he did for all the, did everything for the people that did everything. So and he only had, the only one in Jellingen, and there were about, oh I guess twelve, four, twenty, no, two thousand people. Yeah, that's what they think. Yes, that's what they had. And so they had everything there.

In the English utterances the frequency of empty speech (*all the things, everything* etc.) and unintended anacolutha (*...did for all the...he only had...*) is noticeable. What is absent in her English in contrast to her German utterances are instances of what could be called codemixing. And yet, it seems Mrs. T.'s problems are not so much related to the lack of linguistic or pragmatic constraints but are due rather to a lack of German words or result from the experience that the German words are not readily available. Thus a different interpretation of Mrs. T.'s linguistic behavior is more plausible. In her third conversation (recorded in 2002) she shows that she is aware of her own problems in remembering things or in expressing herself. When asked about her father's name she responds in the following way:

> N: Was war der Name Ihres Vaters? (*What was your father's name?*)
> **T: Meine ist Txxxx. (Mine (in German it should be *Mein Vater heisst...*) *is T.*)**
> N: Und Ihr Vater? (*And your father?*)
> T: Gosh. I think it is...what's the matter with me? That thing is so out.

An interpretation of code-switching or codemixing in terms of a violation of linguistic or pragmatic constraints does not really apply here as the partial non-availability of the linguistic code (German) creates a situation where such implicit rules of constraints can not be applied. She is in a situation that is more comparable to a second language learner who does not have the necessary code to express him- or herself and tries hard to make up for it. The following extract indicates that it is not only vocabulary that causes problems but also a combination of lexis and grammar:

N: . . . . Do you have brothers and sisters? Haben Sie einen Bruder oder eine Schwester? (*Do you have a brother or a sister?*)
T: I have one Schwester and two sisters, two sister.
N: Oh, I see. They are still, leben die noch? Ihr Bruder und Ihre Schwester, leben die noch? (*Are they still alive? Your brother and your sister, do they still live?*)
T: They are. Ja, they are all everyone, but they are not all, all over.

The use of "Schwester" and "sister" in the same utterance could be called codemixing. However, as in the case of the lexical problems that were analysed above, a different interpretation is more in line with the facts if the linguistic behaviour is analysed in greater detail. It appears that the uncertainty about how to express the German plural in "Schwester" – it would be "Schwestern" – and in the word "Bruder" (*brother*), which was actually on her mind as her later utterances reveal – the German plural would be "Brueder" with umlaut – triggers the English "and two sisters, two sister" and so she is able to overcome the difficulties of plural formation and the word-finding process with regard to the German word "Bruder". Again it becomes obvious that this kind of linguistic behaviour is more akin to processes in second-language learning.

What can also be observed in these conversations is that Mrs. T. understands more German utterances than she produces herself. And when she expresses herself in German she hesitates a lot. This is to be expected as both native and second-language speakers who have not used their mother tongue or their second language for a longer period of time become more hesitant in their speech and use code-switching or code mixing more frequently or they respond in their dominant language if the situation allows them to do this. In second-language learning situations, tandem learning makes use of this socio-pragmatic strategy (Byram 2000: 595–7).

N: Sind Sie irgendwann einmal nach Israel gegangen? (*Have you ever been to Israel?*)
T: In where?
N: Waren Sie einmal in Israel? (*Have you ever been in Israel?*)
T: In, in where?
N: In Israel.
T: In Israel?

N: Ja, waren Sie einmal in Israel? In Jerusalem, oder Tel Aviv? (*Yes, have you ever been in Israel? In Jerusalem, or in Tel Aviv?*)

T: Yes, I believe, I think there was once, yeah, but I can't honestly, I don't really anymore too much about that but I know I was.

Summing up Mrs. T.'s linguistic behaviour, it is safe to say that her change from English to German is more motivated by her wish to accommodate her speech to her interlocutor than to a lack of restraints. And if she is unable to retrieve the words she needs to express herself in German, she is more likely to turn to English as her more dominant language than to use empty phrases and in this way she tries to make up for the lack of words or grammatical problems. In her English she compensates for her lack of words by using empty phrases or anacolutha rather than by using German words. In the context of her linguistic community this linguistic behaviour is in line with the socio-pragmatic norms.

There is further proof of this in her speech. In the following extract Mrs. T. accommodates her speech to her interlocutor by taking up a German word (*eine Volksschule*) and using it in a code-switch in her English statement either because there is no equivalent term in English or because she wants to emphasize a particular point or a specific topic (cf. Baker 2001: 104 ff). It can be safely claimed that this behaviour is not an instance of code mixing:

N: Can you still remember the school that you went to when you were in Germany? Did you go to eine Volksschule (*a people's/elementary school*)?

T: Oh yes, first in the Volksschule, and then you were in

N: Excuse me.

T: in

N: Waren Sie in einem Gymnasium? (*Did you go to a grammar school?*)

T: in

N: oder eine Mittelschule? (*or a middle school?*)

T: Yeah.

N: Eine Mittelschule.

T: Yeah, that's right.

N: Ah ja. Und in der Mittelschule haben Sie auch Englisch gelernt, oder? (*Yeah. And in the middle school you also learnt English, right?*)

T: Yeah, oh yeah, definitely.

These extracts demonstrate that Mrs. T. is a personality who can still interact in a very creative way using the two languages at her disposal. The ritual of leave-taking finally underlines the pragmatic skills that Mrs. T. activates and employs successfully at a very late stage of AD in spite of problems she has with her linguistic code:

N: Okay, thanks very much for talking to me, been enjoying it.
T: That's, I'm sure
N: and I hope you will be fine, and we can talk again.
T: Aha.
N: Thanks very much, bye-bye. Auf Wiedersehen. (*Good-bye*)
T: That was very nice.
N: Yeah, I enjoyed it. Thanks very much, I appreciate it.
T: That would be, you, c . . . come to me some time soon again.
N: Yeah, right I will, thank you.
T: This must be very nice.
N: Yeah. Thanks very much. I appreciate it. Bye-bye.
T: All right now, where do I go?

The way Mrs. T. interacts in the leave-taking process confirms that the encounter with a German interlocutor and the exposure to her original mother tongue, German, has been a satisfying experience for her, socially as well as culturally and linguistically.

A native speaker of English with some experience of German

In the study of Alzheimer's discourse two other AD residents were involved in an informal second-language training experience which was developed in the context of conversations and carried out over a period of almost two years. The first eight months of this training were done on a more or less weekly basis during conversations of approximately 10 to 15 minutes of length each. This discussion will focus exclusively on Mr. RW [Robbie Walters]. Mr. RW comes from a family with a German background. It appears that he is a third-generation immigrant; he says his grandparents emigrated to the USA from Germany. There are, however, no indications that German was still spoken in the family when Mr. RW was a child. The German words that he remembers go back to the time when he was a soldier in World War II. As a radio or master sergeant in the Atlantic he picked up the German words that are still at his disposal. At school he must have learnt some Latin, but no German. Mr. RW's English discourse is coherent in most of his utterances, which is fortunate, as much depends on the discourse partner's skill to make inferences about possible connections (Johnstone 2002: 101, 161ff). The following exchange is typical:

N: If you had a choice, what would be your first preference as a golfer?
RW: I would like to ride and take the golf cart. I would take the cart and ride around some. It's a lot more comfortable.

In planning the AD discourse project a central issue was the question if and under what conditions it would be possible to get AD patients involved in second-language training phases so as to recover or newly develop specific components of their communicative competences. The lexical component of linguistic competence needs to be analyzed, because it is part of the declarative knowledge base and it also contains elements of procedural knowledge in so far as lexis is connected with grammar and phonology. As Paradis (2004) points out, individuals with AD have "explicit knowledge" impaired and it is their declarative memory that is affected first (Paradis 2004: 2). In line with this finding it has been observed that Alzheimer patients lose access to their lexical base and make up for the loss by using empty words (Almor *et al.* 1999: 203; Kempler *et al.* 1995). And if a second language is involved, the words of the second language are affected in the first place. As Mrs. T. demonstrates, it is important in an immigrant population to check which of a person's two languages is the dominant, rather than the native language. It is also important to look at the role the sounds of words may play in Alzheimer speech. Paradis, on the one hand, regards phonology as a part of the implicit linguistic competences and, on the other hand, points out that "the sounds and lexical meanings of words are consciously known" (Ibid.).

The effort at second-language training emphasized the process of recovering and developing elements of the German language that the male AD speaker in the project could cope with, on the basis of his current level of L2 competence. His German can be defined in the following way: It is below the lowest level of generative language use (Common European Framework Level A 1), which means that he can only master certain isolated words or fixed phrases. However, fixed phrases and formulas are elements at generative language levels, too (see The Council of Europe 2004: Common European Framework Level A 2). They have an important function in the L2 acquisition and learning process (Ellis 1994: 84ff). So it is reasonable to select a limited number of lexical items and phrases and concentrate on them over a certain period of time. For the training phases in German, fixed phrases were selected, such as those occurring in greeting and leave-taking routines and also clusters of coordinates (cf. Aitchison 1994: 89) such as the days of the week, the months and seasons of the year, numbers and colors.

The training was planned in such a way that it was part of a cycle of conversations that were carried out at regular intervals over a period of time. The AD speaker was not supposed to get the impression that he was being taught anything, in order to create a non-threatening atmosphere for learning and to keep the flow of speech as natural as possible. At one point this nevertheless became an issue in a conversation with Mr. RW; the following extract from 2001 shows how informal the situation was in spite of the training, and how the topic of school could be taken up naturally:

> N: . . . Wie gehts heute morgen? Wie gehts? How are you? Gut? Gehts gut? Gehts gut?
> Are you fine? <*These phrases are part of the intended training*>
> RW: In school?
> N: No, here. You are not at school. You are here this morning enjoying your time.
> RW: Oh, I just thought I heard the question, but I didn't know.
> N: Oh, I see. It reminded you of school, did it? That's funny. Can you tell me what school it was that you went to? [2001]

In this situation the interlocutor takes Mr. RW's statement seriously, accepts that he is reminded of school and uses his idea to change the topic of the conversation with a shift away from the training situation to the topic of school.

As the linguistic focus of the training phases was on developing the AD patients' linguistic declarative knowledge – fixed phrases and clusters/sets of semantic coordinates such as the days of the week – several memorizing strategies were used to make the language items more accessible. The Alzheimer patients were encouraged with repetition and to form visual links with a calendar or pictures to talk about dates (numbers), days, months, and seasons or with playing cards, where the numbers on the cards were pointed at in German. Thus a typical training phase within a conversation would proceed in the following way:

RW, 2000: first week of training

<Training was always embedded within a conversational frame to support topic change>

N: Good morning, "Robbie"! Nice seeing you again!

RW: Thank you! I appreciate that! It's always nice to talk to you!

<Here the training phase starts with a focus on greeting and then numbers in German>

N: Wie geht's! How are you? Wie geht's? (*How are you doing?*)

RW: I'm fine.

N: Or in German we can say . . .?

RW: Gut?

N: That's it!!! Wonderful!

RW: Yes.

N: All right! Do you remember we talked about numbers last time? You remember?

RW: Eins, zwei, drei, vier, fünf, sechs. (*German numbers from 1 to 6*)

N: Sieben. (*7*)

RW: Sieben.

N: Acht. (*8*)

RW: Acht.

N: Neun. (*9*)

RW: Neun.

N: Zehn. Zehn. Ten.

RW: What? Oh, Zehn. Ten.

N: Right. Ten. . . .

<Subsequent training proceeds: it may have a new objective, or it is interrupted and conversation resumes. Note how RW self-corrects, asks for repair, clarification or repetition, and finally produces the correct sequence, though he is initially dependent on N's initiation of the words.>

[*N continues*] Today I want to talk about the days of the week. You remember Sunday?

That is "Sonntag". You remember?

RW: No.

N: Then I'll tell you. Sonntag: that is Sunday.

RW: Sunday.

N: Sonntag.

RW: Sonntag: Sunday.

N: Yes. Montag like the moon.

RW: Moon. Montag.

N: Sonntag, Montag.
RW: Uh huh.
N: Dienstag: That's Tuesday.
RW: Huh?
N: Dienstag. Sonntag, Montag, Dienstag.
RW: Montag?
N: Montag is Monday. Tuesday is Dienstag.
RW: Dienstag. Sssssonntag . . .
N: Sonntag, Montag.
RW: Montag.
N: Dienstag.
RW: Dienstag.
N: Right! Once more! Sonntag . . .
Look at my lips: Sonntag . . .
That's it! Good! Great! Sonntag, Montag, Dienstag. You're good at remembering things!
And you can count from one to ten. That's great!

<Occasionally a second phase of repetition focuses on an earlier part of the training>

[*N continues*:] Can we do the counting once more?
RW: Eins, zwei, drei, vier, fünf, sechs, sieben, acht (*German numbers from 1 to 8*). . . .

<Return to the conversational frame – in this case initiated by the AD speaker>

RW: I got along with the German people real well.
N: That's right. You told me that was a long way back. Yes. When you worked in your government office, you didn't have to speak German?
RW: No. See, that kept me from having the experience I would have liked to have had with them. Because, like I say, I liked the German people.

This extract from AD speech demonstrates that even at moderate stages of AD it is possible to involve an AD speaker in second-language training quite naturally. The comments made by the AD speakers on their experience in this training indicate that they felt their self-esteem was enhanced: they expressed their views in very positive terms and they were looking forward to further experiences of this kind.

Apart from the affective side of this experience (cf. Arnold 1999) it is important to investigate if there was an impact of the training on the language abilities of the AD speakers and it is also necessary to find out if there were any effects over a longer period of time. In order to be able to judge possible training influences, the speech data of the AD speakers that were collected over a period of almost two years are analysed with a special emphasis on the speech behavior of the AD patients at various stages during this period. A test-based research approach would not be in line with the informal nature of the second-language training in this research project and it would not yield results that would reflect what the AD patients were able to do.

The L2 linguistic competence of Mr. RW at the point immediately before the first training phase can be observed in the discourse sample recorded one week prior to the training detailed above. Two weeks before the training phase, there had been a brief exchange (greeting) in German.

RW: Yeah I know a few words like ummph
N: (*Laughed*)
RW: **Heinz sprechen** (*Heinz <German first name> speak*)
N: (*Laughed*) an' what was it for, Guten, uh, good morning, what do you say in German?
RW: Guten Morgen....
N: Excellent can you, uh, do you remember
RW: = [*Unclear and overlapped*]=
N: the numbers for one
RW: ein zwei drei .. uh . . . (*1 2 3*)
N: Yeah that's right, that's right
RW: umh
N: vier fünf (*4 5*)
RW: umh
N: sechs sechs sieben (*6 6 7*). . . .;
<Here the training phase begins>
N: Yeah that's right uh well we can count again one er = eins, zwei, drei, vier, fünf, sechs, sieben (*1 2 3 4 5 6 7*)
RW: ein zwei drei vier fünf sechs sieben - huh?....
<After this first phase the numbers are counted down to one and up to 10 twice>

As can be observed in this text in comparison with the text above, recorded the following week, Mr RW's performance is very restricted here in terms of number of words that are accessible and with regard to the articulation of the numbers. A week after the first training, his performance has improved considerably: He can use more words without help and repeats numbers with greater ease and with a German pronunciation. It is also noticeable that he considers his German words as a part of his own ability and he is proud of it as his final statement after the training phase indicates. The reasons for this improvement can ultimately not be judged as it might be a process of recovering and making available for use what had been acquired at an earlier time in the life of the AD person or it could be a new learning process. The question now is if this short-term training effect in Mr RW's case might last for a longer period of time. For this reason Mr RW's second language behavior is highlighted after a longer time of training and after an interval without any conversation in German.

The training continued for three months, then there was an interval of four weeks, before the conversations were resumed. The following discourse was recorded four months later, in spring 2000.

RW after an interval of four weeks without training

> RW: All of these are March? *<Looking at a calendar>*
> N: All March. That's 1,2,3,4,5,6,7,8,9,10. Was it March the fifteenth? Or was it more like March the fourth, your birthday?
> RW: My birthday? . . .
> N: Ok. Actually, do you remember some of the numbers that we talked about in German? Eins, zwei, drei.
> **RW: Eins, zwei, drei, vier, fünf, sechs.** (*Numbers from 1 to 6*)
> N: Sieben.
> **RW: Sieben.** (*7*)
> N: Acht.
> **RW: Acht.** (*8*)
> N: Neun. (*9*)
> **RW: Novem.**
> N: Neun. Zehn. Ten zehn. Good! (*10*)
> N: We also said the names of some of the months in German, too. In German, we would say Januar (*January*).
> **RW: Januar.**
> N: Then Febuar (*February*).
> **RW:** February and then my month, March. [*March is his birthday month*]

In the first exchange after a period without any contact with German, Mr. RW is still more fluent than before the training phases began; he can remember the numbers that had been practiced although he has a slight problem with number 9, where his knowledge of Latin seems to interfere with German. He is, however, able to correct himself with help. A finding like this would be judged to be in line with expectations in the case of second-language learners in a classroom, although the number of words would be considerably greater among L2 speakers without AD. As for the days of the week or the months of the year Mr RW's performance never reached the same level of independence as in the case of the numbers in German. So his performance here is a fair representation of his knowledge of days and months that he had reached, which implies that his limited skill in this field remained at the level reached. However his independent use of English days and months was also restricted in contrast to his knowledge of English numbers. This may be a hint that his ability to recover or to learn words in German is related to his respective L1 competence.

Following another training phase, his performance is at its best two weeks later:

RW: Well, I like this fall picture in area and time.
N: I see.
RW: That time frame and all.
N: That's interesting. Can you still remember some of the numbers that we used for counting?
RW: Eins, zwei, drei, vier, fünf, sechs, sieben, acht, neun, zehn! *(1–10)*
N: Excellent! You are very good at that! I like it! Do you also remember the days of the week that we mentioned in German? Like the first day of the week in English? Sunday?
RW: What?
N: Sunday is the first day of the week in English.
RW: Sunday?
N: Yes. In German this was ...? Do you still remember? It also started with an "S."
RW: Sunday, Monday ... I don't know.
N: If I said them in German, would you remember? Sonntag, Montag, Dienstag. Any of them that you remember?
RW: No, I don't, not too well.
N: But you may remember Sonntag?
RW: Huh?
N: Sonntag, that's Sunday. Sonntag.
.... So we say Sonntag, Montag, Dienstag. Can you say that?
RW: Sundag? *<a phonetic approximation>*
N: Sonntag. Montag.
RW: Sonntag, Montag.

N: And Dienstag.
RW: Dienstag.
N: That's right. That's my German language.
RW: Well, you do alright.

Note at the end of this text extract Mr. RW comments on the interlocutor's performance, which suggests that he felt he was accepted as an equal partner in discourse.

The long-term effect of the German training experience was again documented after an even longer interval without any German language contact. Mr. RW did not have any contact with German for four months. After such a period without any language contact the L2 competence of a learner without AD would be expected to be somewhat diminished in comparison with the language level reached at the end of a learning period.

RW six months later, after an interval of four months without training

N: Mr. Robbie! Nice to meet you again! Do you remember me?
RW: Yes....
RW: ...Mostly, I've been bored. That's the only problem.
N: That's true, yes. Do you still remember some of the German? We counted the numbers, didn't we?
RW: Eins, zwei, drei, vier, fünf... (*1–5*)
N: Oh! Wonderful! You're still the language expert, aren't you? Eins, zwei, drei...
RW: vier, fünf, acht, neun... (*4, 5, 8, 9*)
N: Zehn!
RW: Zehn. (*10*)
N: Yes! That's right! We said...what did we say for, "Good morning?" We said, "Guten Morgen." Guten morgen, QXXX! Good morning, QXXX. Guten Morgen, yes.
RW: Guten Morgen.
N: Yes! Guten Morgen!...Here, wait a moment...this was a specific month. Do you remember the month? Can you see it here?
RW: March: my birthday!
<Mr. RW sticks to English when asked to talk about the months>

As can be detected here, Mr. RW's German performance six months later and after an interval of four months without any training is comparable to the situation after the last shorter interval without German contact. His L2 performance is slightly reduced, although it clearly reveals that the German words are still available in ways similar to the time at the end of the last training experience. It is also consistent in the sense that he is focused on German numbers, whereas he has problems with days and months.

Discussion and summary of the findings

An ethnographic research approach was used to collect and to analyse data about the discourse of two Alzheimer patients, based on theoretical considerations that function as guidelines for the analysis of the empirical data. Thus the research questions relate to various aspects of AD speech behavior from a theoretical perspective and they focus on the specific background and the peculiar speech behaviour of each patient individually. In contrast to a quantitative approach, this method sets great store by a more comprehensive analysis of individual speech behaviour, which makes it hard to draw general conclusions. However, a greater knowledge of the variables and processes that are involved in AD discourse can lead to a deeper understanding and can raise new questions.

(1) The first research question directed attention to the variables that are connected with the unimpaired interlocutor in AD discourse. The analysis in this chapter refers to this question only implicitly. What would be needed to address this question in detail is an investigation of AD discourse that also includes the other members of the research team as interlocutors and looks at topic choice. However, the analysis of AD second-language behaviour here yields results that are relevant to further research. The fact that the interlocutor in this study is a native speaker of German and a second-language speaker of English was a trigger for the AD speakers to use the second-language phrases and words at their disposal. They did not speak German with other members of the research team and did not show signs of code-mixing or code-switching when they talked English. In addition, the AD patients considered their German interlocutor as a partner in discourse and tried hard to adapt their limited L2 speech to his speech in German. The reverse is true for the German interlocutor.

Moreover, there was no problem that could be related to the different accent of the interlocutor in English (a mostly transatlantic English accent). The AD patients were willing to accept the interlocutor as a personality. Gender was an issue in the sense that the male AD discourse partner selected or expanded topics in English that differed from topics in his discourse with the female interlocutors.

(2) The second research question is focused on the impact of the language background of the AD speakers on the interaction in their conversations. One is a first-generation American lady whose mother tongue was German before she left Germany at the age of 18, at a time when her L1 had reached an adult level of competence. In her case it is essential to realize that English has become her dominant language, as her interaction reveals. She converses primarily in English that is like other AD speakers: When an English word or phrase is not available or accessible her speech tends to be semantically empty or the number of unfinished statements increases. When conversing with the native speaker of German she occasionally also makes use of strategic code-switching (cf. de Picciotto & Friedland 2001). There are noticeable differences in her use of code-mixing and code-switching depending on whether she speaks English or German. It is only in her German statements that she switches or mixes codes and in each case it is English she uses, to make up for the lack in her German.

The other AD speaker has a very limited command of German phrases and words. He does not have a creative competence in German although he picked up his German in terms of naturalistic language acquisition (Ellis 1994: 17). His interactive behaviour in German language training phases over an extended period of almost two years – with interruptions of up to several months – reveals that he tries hard to participate in German language activities.

For studies that investigate bilingualism in the field of AD it is of great importance to describe the linguistic and cultural background of the subjects in detail (Hyltenstam & Stroud 1993). Studies that focus on language deficits, e.g. using naming tasks to analyse correctness levels (cf. Goral 2004: 33), may be unaware of the peculiar variables that relate to the language background of the AD speakers.

(3) The question of whether AD speakers can access L2 vocabulary can be answered positively. However, the way words are accessed has a tremendous influence on the results. In a semantic priming or naming task (cf. Goral 2004; Phillips *et al.* 2004) the subjects cannot go through a process that gradually paves the way for a less accessible language to emerge, as could be observed in the case of the first-generation American in this study. If an AD speaker needs to be involved in a process of co-construction (cf. Davis & Moore 2003: 122) the way he or she accesses L2 vocabulary has to be taken into account.

(4) A special interest of this study is whether AD speakers can recover or newly develop elements of an L2 by being involved in a second

language training experience. As far as the ability to recover words and phrases is concerned, the analysis of the speech behaviour indicates that two AD speakers are able to recover and retrieve words. If the range of words is taken into account and if it becomes obvious that AD speakers more or less remained on a plateau throughout the training experience, the theoretical considerations are not refuted in spite of the fact that they may have to be somewhat modified to be more inclusive.

(5) The question of which language in a bilingual AD speaker is affected first or which components of the languages are more liable to language loss is of great social significance in a society with an immigrant population. The case of the first-generation American in this study shows that there is a negative impact on both the L1 and the L2 of the AD speaker. Also, the less dominant language is affected by AD more than the dominant language. Or in other words, it does not seem to be so important which language is the first or the second-language acquired or learnt. This finding is in line with some of the findings that Goral (2004, 38) summarizes in her article. She detects a clear influence of the research methods on the results. Our investigation supports this judgment. However, we stress that this investigation has focused on what Alzheimer's speakers are still able to do in a process of accommodation that involves an AD speaker and a competent partner in AD discourse.

References

Aitchison, J. (1994) *Words in the Mind: An Introduction to the Mental Lexicon*. Oxford: Blackwell.

Almor, A., Kempler, D., MacDonald, M.C., Andersen, E.S. & Tyler, L.K. (1999) "Why do Alzheimer patients have difficulty with pronouns? Working memory, semantics, and reference in comprehension and production in Alzheimer's disease." *Brain and Language* 67: 202–27.

Arnold, J. (1999) (ed.) *Affect in Language Learning*. Cambridge: Cambridge University Press.

Baker, C. (2001) *Foundations of Bilingual Education and Bilingualism*. Third Edition. Clevedon: Multilingual Matters.

Bayles, K., Tomoeda, C. & Trosset, M. (1992) "Relation of linguistic communication abilities of Alzheimer's patients to stage of disease". *Brain and Language* 71: 454–72.

Blumstein, Sheila E. (1988) "Neurolinguistics: an Overview of Language–Brain Relations in Aphasia." In F.J. Newmeyer (ed.) *Linguistics: The Cambridge Survey. Volume III: Language: Psychological and Biological Aspects*. Cambridge: Cambridge University Press, 210–36.

Byram, M. (2000) *Routledge Encyclopedia of Language Teaching and Learning*. London and NY: Routledge.

Caplan, D. (1988) "The Biological Basis for Language". In F.J. Newmeyer (ed.) *Linguistics: The Cambridge Survey. Volume III: Language: Psychological and Biological Aspects*. Cambridge: Cambridge University Press, 237–55.

Chomsky, N. (1980) *Rules and Representations*. Oxford: Basil Blackwell.

Coupland, N. & Jaworski, A. (eds.) (1997) *Sociolinguistics: A Reader and Coursebook*. London: Macmillan.

Cummings, J.L. (1992) "Dementia. The failing brain", *Lancet* 345: 1481–4.

Cummings, J.L., Benson, F., Hill, M.A. & Read, S. (1985) "Aphasia in dementia of the Alzheimer type." *Neurology* 35, issue 3: 394–7 <http://www.neurology.org/cgi/content/abstract/35/3/394.>

Davis, B. & Moore, L. (2003) " 'Tho' Much Is Taken, Much Abides': Research on Retention in Alzheimer's Discourse". In J. Rymarczyk and H. Haudeck (eds.) *In Search of the Active Learner*. Frankfurt am Main: Peter Lang, 115–27.

De Picciotto, J. & Friedland, D. (2001) "Verbal fluency in elderly bilingual speakers: normative data and preliminary application to Alzheimer's disease." *Folia Phoniatrica et Logopaedica* 53, No. 3: 145–52.

Ellis, Rod (1994) *The Study of Second Language Acquisition*. Oxford: Oxford University Press.

Elman, J. (2003) "Development: it's about time". *Developmental Science* 6: 4, 430–3.

Elman, J, Bates, E., Johnson, M., Karmiloff-Smith, A., Oarisi, D. & Plunkett, K. (1996) *Rethinking Innateness: A Connectionist Perspective on Development*. Cambridge, MA: MIT Press/Bradford Books.

Friedland, D. & Miller, N. (1999) "Language mixing in bilingual speakers with Alzheimer's dementia: a conversation approach." *Aphasiology* 13: 427–44.

Giles, H. & Powesland, P. (1997) "Accommodation Theory". In N. Coupland & A. Jaworski (eds.) *Sociolinguistics. A Reader and Coursebook*. London: Macmillan, 232–39.

Goral, M. (2004) "First-language decline in healthy aging: implications for attrition in bilingualism". *Journal of Neurolinguistics* 17: 31–52.

Grimm, H. (ed.) (2000) "*Sprachentwicklung*". Göttingen, Bern: Hogrefe Verlag für Psychologie.

Hamilton, H. (1994) *Conversations with an Alzheimer's Patient*. Oxford: Oxford University Press.

Heindel, W.C., Cahn, D.A. & Salmon, D.P. (1997) "Non-associative lexical priming is impaired in Alzheimer's disease". *Neuropsychologia* 35: 1365–71.

Hyltenstam, K. & Stroud, C. (1993) "Second language regression in Alzheimer's dementia." In K. Hyltenstam & Viberg, Å. (eds.), *Progression and Regression in Language*. Cambridge: Cambridge University Press: 222–42.

Kempler, D. (1995) "Language Changes in Dementia of the Alzheimer Type". In R. Lubinski (ed.) *Dementia and Communication*. Philadelphia: B. Decker, 98–114.

Johnstone, B. (2002) *Discourse Analysis*. Oxford: Basil Blackwell.

Meguro, K., Lh Senaha, M., Caramelli, P., Ishizaki, J., Chubacci, R.Y.S., Meguro, M., Ambo, H., Nitrini, R. & Yamadori, A. (2003) "Language deterioration in four Japanese-Portuguese bilingual patients with Alzheimer's disease: a transcultural study of Japanese elderly immigrants in Brazil." *Psychogeriatrics* 3, Issue 2: 63.

Mendez, M.F., Perryman, K.M., Pontón, M.O. & Cummings, J.L. (1999) "Bilingualism and dementia". *Journal of Neuropsychiatry and Clinical Neuroscience* 11: 411–12.

Muñoz, M.L., Marquardt, T.P. & Copeland, G. (1999) "A comparison of the codeswitching patterns of aphasic and neurologically normal bilingual speakers of English and Spanish". *Brain and language* 66: 249–74.

Newmeyer, F.J. (ed.) (1988a) *Linguistics: The Cambridge Survey. Volume III: Language: Psychological and Biological Aspects.* Cambridge: Cambridge University Press.

Newmeyer, F.J. (ed.) (1988b) *Linguistics: The Cambridge Survey. Volume IV: Language: The Socio-cultural Context.* Cambridge: Cambridge University Press.

Obler, L., de Santi, S. & Goldberger, J. (1991) "Bilingual Dementia: Pragmatic Breakdown". In Lubinski, R. (ed.) *Dementia and Communication.* Philadelphia: B.C. Decker, 133–9.

Paradis, Michel (2004) "Neurolinguistics of bilingualism and the teaching of languages". 27.05.2004 <http://semioticon.com/virtuals/talks/paradis_txt.htm>

Phillips, N.A., Segalowitz, N., O'Brien, I. & Yamasaki, N. (2004) "Semantic priming in a first and second language: evidence from reaction time variability and event-related brain potentials." *Journal of Neurolinguistics,* 17: 237–62.

Ramanathan, V. (1997) *Alzheimer Discourse: Some Sociolinguistic Dimensions.* Mahwah, NJ: Erlbaum.

Saville-Troike, M. (1997) "The Ethnographic Analysis of Communicative Events". In N. Coupland & A. Jaworski (eds.) *Sociolinguistics: A Reader and Coursebook.* London: Macmillan: 126–44.

Slobin, D.J. (1979) *Psycholinguistics.* Glenview, Ill.: Scott, Foresman (2nd edn).

Spolsky, B. (1988) "Bilingualism". In F.J. Newmeyer. (ed.) *Linguistics: The Cambridge Survey. Volume IV: Language: The Socio-cultural Context.* Cambridge: Cambridge University Press: 100–18.

The Council of Europe (2004) *The Common European Framework of Reference for Languages.* Cambridge: Cambridge University Press.

Weinert, S. (2000) "Beziehungen zwischen Sprach- und Denkentwicklung". In H. Grimm (ed.) (2000). *Sprachentwicklung.* Göttingen, Bern: Hogrefe Verlag für Psychologie: 311–61.

Whorf, B.L. (1956) *Language, Thought, and Reality: Selected Writing of Benjamin Lee Whorf.* Edited and introduced by J.B. Carroll. Cambridge, MA: MIT Press.

Wray, A. & Perkins, M. (2000) "The function of formulaic language: an integrated model." *Language and Communication* 20: 1–28.

7
So, You had Two Sisters, Right? Functions for Discourse Markers in Alzheimer's Talk

Boyd H. Davis

Naturally occurring conversation with Alzheimer's speakers, including embedded or co-constructed narrative, can be used to support or augment clinical findings on features of Alzheimer's discourse, which are as highly variable as is the disease. For example, not only can speakers with moderate to moderately severe Alzheimer's disease maintain some level of politeness (Sabat & Collins 1999; Temple *et al.* 1999; Rhys *et al.* 2000); and interaction (Hamilton 1994a, 1994b; Ramanathan 1997), their pragmatic skills can sustain more fine-tuned analysis, particularly as those skills sustain or simulate fluency. Ellis (1996) comments that problems Alzheimer's speakers have in organizing and concentrating information are both communicative and cognitive. In the early stages, as grammatical modes of processing deteriorate for whatever reason, those "cohesion ties that structure topicality will begin to fade"; in later stages, more severe problems with maintaining topic will surface (see for example, Kempler 1995). Alzheimer's speakers typically depend on lexical cohesion, or word-based means of holding the elements of a sentence together, as opposed to grammatical because, adds Ellis, lexical cohesion relies on "meaning," not on grammatical structures. They lose the ability to "ground" the discourse for the hearer, or to organize meaning with thematic information via pronouns (see Almor *et al.* 1999)

However, there are other ways to connect pieces of discourse: the so-called "little words" such as *oh* and *so* do quite nicely (Fox Tree and Schrock 2002; cf. Schrock & Fox Tree 2000). In this discussion, we examine *oh, well* and *so* in the speech of three women whose disease is at different levels of severity in order to inventory the range of functions of retained discourse markers in naturally occurring conversations with speakers diagnosed as having Alzheimer's disease. Current notions of conversational competence and models for politeness, grounding, audience design or evaluation in narrative can be examined in light of what these speakers retain. Careful description and initial analyses of discourse features retained by Alzheimer's speakers might, for example, contribute to multidisciplinary discussions of isomorphism, more generally called the regression hypothesis, or whether the pragmatic component of language, however defined, is modular (Kasher 1998; cf. Perkins 1998). Pragmatic features of discourse are not, note Stemmer & Schoenle (2000: 10) "subserved by the same brain systems involved in producing aphasia." They add that exploration of discourse by neurologically impaired persons is a fairly recent phenomenon: we do not know what we will find when we include speakers whose dementia-based communicative problems are, as with Alzheimer's, diffused across several areas of the brain. Such description and analysis is warranted from multiple disciplines.

With Alzheimer's speakers, it may be useful to consider some of the "little words" as acting very like formulaic discourse in the ways they suggest fluency and support social interaction. Formulaic speech, as noted by Norrick (2000: 48) is two or more words, which includes phrasal verbs, binomials ("free and easy"), recurrent collocations, such as "live it up," and proverbs. According to de Santi's study of idioms and formulaic language in Alzheimer's discourse, "those sentences which are highly familiar and over-learned remain relatively spared in Alzheimer's disease, whereas those sentences which are new and unfamiliar are difficult to repeat without error" (de Santi 1993: 139). Either the idiom becomes "automatic" and preserved in aging, or it is stored as chunks which means the sentences require less processing (ibid.: 141). However, comparison with control groups of normally-aging older people, as well as with unimpaired youth (Erman 2001; Macaulay 2002; Jucker 1993; Jucker *et al.* 2003) suggests that "language does not become more automatic with age, but the aging adult might perform at its (*sic*) best when dealing with familiar/over-learned language" (de Santi 1993: 159). The uses of initial *so* and *well*, for example, pattern in ways similar to other chunks of functional expressions, lexical or grammatical collocations, and frozen or formulaic discourse (Wray & Perkins 2000; Wray 2002), in the discourse of our speakers.

In what follows, we first list functions for three discourse markers in the speech of three women at different stages of Alzheimer's disease, and in different social settings. Reviewing issues of reception and production has led us to look more closely at interactions containing indirect questions, and at participant roles in the conversations. Here, we compare our findings with Fuller's review (Fuller 2003) of the impact of speaker roles on the choice for discourse markers such as *well*. Finally, we will comment on an implication for interventions designed to enhance communication.

Oh and *well* in the speech of three Alzheimer's speakers

If it is hard to define a discourse marker in non-impaired speech, it is equally so in the speech of those who are cognitively impaired by Alzheimer's. Kim *et al.* (2000) indicate the range of opinion on discourse markers, noting they "have as many different descriptions as people describing them". While Schiffrin (1987) remains a necessary grounding in the identification and explanation of functions for discourse markers, additional scholars have augmented her discussion. Kim *et al.* (2000) cite, for example, Grosz & Sidner (1986) for whom "discourse markers flag changes in both attentional and intentional state" as opposed to the approach of Mann and Thompson's Rhetorical Structure Theory, where "discourse markers mark rhetorical relations between segments" (Kim *et al.* 2000: 262). Fischer & Drescher (1996) hold that discourse particles "fall on the border" between semantics and pragmatics, and that (p. 854) they "contribute to the micro and macro structure of dialogues" with their frequency of occurrence dependent on the domain in which they occur. Aijmer (2002) notes the multi-functionality and indexicality of discourse particles in her corpus-based study. Following Katzenberger we see discourse markers as having two functions: "as 'procedural tools'.... [to] indicate the discursive frame of the text that is being produced... [and to] instruct addressee(s) how to interpret parts of the text or the text as a whole (Blakemore 1996)" (Katzenberger 2004). For Katzenberger, discourse markers also serve to suggest or cross-reference a genre or domain of text: that is, whether it is narrative, expository, or descriptive text; as well as the position and function of the textual segment the discourse marker prefaces. In that sense, we may speak, for example, of a discourse marker as signaling the evaluation in a narrative.

The discourse markers we summarize below are from several years of unprompted conversations in natural settings with three women. Two of the speakers, Glory Mason and Aileen Copeland, live in different areas at Pleasant Meadows: one in a wing for non-ambulatory residents, an apartment, and the other in an assisted living apartment. Although the conversations with them from 2000 to 2004 occur in natural settings and are unprompted, they nonetheless often sound like interviews, since the conversations, particularly with Mason, include a great many question–answer exchanges from our effort to locate a topic on which we might have a conversation. The third speaker, Ruby Brewster, was recorded at home prior to her death by her daughter-in-law, who shared the transcripts in order to furnish a comparison across social settings (see Brewer, this volume).

Glory Mason has early moderate cognitive impairment; however, her severe deafness and poor vision contribute to her seeing us as strangers, if friendly ones, each time she has met us. Our talk includes a great many questions, answers, and conversational repairs. Living in the same region as Mrs. Mason for most of our lives, we can style shift into a gender-marked variety fairly close to her regional standard, Piedmont North Carolina, though our version might be considered "citified." In the four years of conversations (2000–2004) with Aileen Copeland, she has moved from mild to early moderate impairment. From 2000 to 2003, she has greeted us as familiar acquaintances; in 2004, seeing us only a handful of times, we are probably seen as friendly strangers or chance acquaintances: vaguely familiar, but without the security of shared dialectal features, as she is from another area of the country. Ruth Brewster's conversations were recorded as she moved from moderate to severe impairment in a family situation. The conversations took place as she was surrounded by family, including some of who were with her in an earlier part of her life, and who could shift into something approximating her regional variety of Oklahoma English. We mention regional dialects as having an as-yet unmeasured impact on gender and regional accommodation in conversation with cognitively impaired speakers and which, in addition to presenting lexical variation, may impact both reception and production of intonational contours, amplitude or shifts in emphasis. Future studies of this corpus will examine regional features more closely. In the meantime, we reference Mahendra *et al.* (1999) who note that tests of speech discrimination and phrase repetition by both unimpaired and Alzheimer's speakers were adversely affected by unfamiliar accent.

According to Jucker, *well* typically signals that "the most immediately accessible context is not the most relevant one" to use in interpreting the upcoming utterance (1993); in his review of its use across historical periods of English, he notes (1993) that it has four uses in Modern English:

as a qualifier it may preface a reply which is only a partial answer to a question;
as a face-threat mitigator, it may preface a disagreement;

as a hesitation marker it may bridge interactional silence; and
as a frame it may be used to introduce a new topic or to preface direct reported speech.

Van de Craen notes (2000) that non-Standard Dutch *alle/allez* functions in similar ways. In their analysis of how people use discourse markers in task-oriented dialog, based on the annotated TRAINS corpus, Byron and Heeman insert a reminder of the importance of interactional genre when they note that *well* in the TRAINS corpus is "typically used to correct a misconception or to suggest an alternative plan" (1997: 3). *Well* can also signal an unexpected response: we add that this could be a signal by the speaker that what she has just heard was unexpected, or that what she will say next could be different from what she thinks the partner expects to hear.

The marker *oh* often signals a change of state for the speaker (Heritage 1984): that is, the speaker has undergone a change in her state of information, orientation or awareness. Katayama (2001) claims that the particle can occur as a mutual product of the speaker and the interlocutor, particularly when signaling something from the affective domain, as with "emphatic oh": Aijmer (1987) calls this use an *evincive*. The interactional nature of *oh* is suggested by the TRAINS corpus, where Byron and Heeman find that *oh* can mark a repair or it can signal that the other speaker has provided new information, or that the previous information was incorrect in some way. All three women present *oh* and *well* with the following functions:

oh in response to new information: change of state
affective reply to indirect or hypothetical solicitation with affiliative response expected; occurs after questions presenting new information (solicitation)
delay device
face-threat mitigator, pointing to contradiction or possible problems
(see e.g. Heritage 1984, 1998a, 1998b; Schiffrin 1987)

well delay device
frame ratification or extension for farewell sequence
face-threat mitigator, often in formulaic phrases; utterance-initial,-medial and-final. When utterance-initial, introduces potential contradiction
topic changer
qualifier for agreement, prefaces incomplete, non-direct, qualified answer
(see e.g. Mueller 2003; Aijmer 2002; Jucker 1993; Jucker *et al*. 2003; Schourup 2001)

Ruby Brewster, Glory Mason and Aileen Copeland continue to try to be good speakers in the sense that they are trying to achieve or to simulate fluency. Clark (2002) notes that good speakers use four strategies to synchronize their delivery of speech with the expectations of their addressees: (1) signal when you want to initiate speech; (2) produce speech as if it were with ideal delivery; (3) signal if you must suspend speech; (4) signal if you must delay speech. For Clark, a particle such as *Well* suggests the first strategy, a collateral signal that one speaker is committed to beginning speaking. Both Brewster and Mason typically use *well* in the same way they use *guess so,* which for each of them is a formulaic phrase that mitigates threats to face. With Brewster, there is a slight but potentially significant difference in her usage as she moves into a later stage of the disease. When she suffered from mild to moderate Alzheimer's, her face-threat mitigators were signals that she was offering contradictory or unexpected information; as she moved into a later stage of the disease, she became more tentative in her signals and ways she hedged her content. It may be, as Rhys *et al.* have recently suggested (2000), that surface tact demands one approach to management of pragmatic features, and politeness as a system requires another: they find that Alzheimer's speakers retain a desire to save their own face though they may not always be capable of showing a concern for the face of the interlocutor. For example, in a conversation with her son CB and daughter-in-law RB, Ruby Brewster uses *well* to signal that she really does not want either choice of sandwiches she has been given and to end the topic

CB: You want bologna or pimiento cheese?
JB: I would have pimiento cheese
RB: *Well,* I don't know anything I like

Compare this with Copeland's *well,* in a similar discussion: here, Copeland is able to suggest a solution to a problem, as opposed to refusing any choice:

LM: Won't that be a problem during all of Lent if you don't like fish?
AC: *Well,* I can have something like an omelette

Conversation and interview

A speaker's fluency may be differently perceived in different contexts. Fuller (2003) compares *you know, like, oh, well, yeah,* and *I mean* for the same speakers across an interview, a narrative and a casual conversation in a study of adult second-language learners. Her findings

> support the claim that the asymmetry of the relationship between the interviewer and interviewee plays a large role in the pattern of use of these DMs, and suggests that *well* may be frequently used as a reception marker. Both *oh* and *well* are used most frequently by hearers who are in a role of listening and responding; speakers who are the narrators (Fuller 2003: 41)

In so many of our conversations, we sound almost like interviewers, bending the conversation to our agenda, with direct questions that go unanswered, and distancing ourselves back into the role of strangers. Caregivers in institutional settings face this cline each day, ranging from strangers every time we meet to close friends or family members. Somewhere in the middle is a continuum for familiar acquaintances ranging from marginally or vaguely familiar to reasonably familiar for name and/or face, and resumption of recognized roles. More work with discourse markers and their production as backchannels, or feedback, may turn out to be highly useful, as we found when we began to examine a third discourse marker, *so*.

So-prefaced statements and indirect (declarative) questions

We found the following functions for *so*-prefaced statements in our data for Mason and Copeland, although they occurred but rarely in their speech:

So-^{1} speaker understands prior turn; shades or highlights topic, may offer new information
So-^{2} speaker returns to main or higher-level context
So-^{3} speaker initiates evaluation component of narrative
As with a *so* indicating "after this, therefore because of this", any of these uses, though particularly So-^{3}, may be conjunctive in nature rather than being discourse markers: see the most recent Annotation Guidelines for Metadata sponsored by the Linguistic Data Consortium [in June, 2004, online at http://www.ldc.upenn.edu/Projects/MDE/]

Instead, we found both Mason and Copeland responding to *so*-prefaced statements by others, particularly when the *so*-statements were indirect or declarative questions, and responding with both additional length and elaboration with new details.

According to Byron & Heeman (1997), *so* provides conclusions and summaries in the TRAINS corpus, although it is typically seen as returning to a main or higher level after a clarification. Copeland's response to *so*-prefaced statements can signal her understanding of a macroproposition begun in small talk and act as a turn-starter, or it can be an indirect request for additional information from the interlocutor. She is able, then, to handle inference and to expect her interlocutor to be aware that she can present content capable of sustaining inferentiality. When we looked more closely at Mason, whose impairment is stronger, we found that she, too, could handle inference in *so*-prefaced declarative questions.

Declarative questions, particularly prefaced by *so*, turn out to be excellent conversation extenders with Alzheimer's speakers in that they provide enough information to support recognition of information, which is less difficult ("more robust") than recall (Bayles 2003). It is difficult to be sure of the role here for prosody, typically defined as speech intonation and emphasis, although intuitively one would expect prosody, syntax and semantics to support each other. Safarova & Swerts (2004) cite research by Geluykens on "question-prone" utterances. As they comment,

> a sentence like "You feel ill" is likely to have an interrogative intent, since one cannot easily make a statement about the inaccessible internal state of another person. On the contrary, "I feel ill" is more statement-prone, as the speaker is not likely to question his/her own feelings.... In follow-up studies using spontaneous speech corpora of Southern British English, Geluykens (1988) found that a majority of declarative questions in his corpus occurred with a fall (57% of the data, with the overall frequency of falls – 64%). On the basis of his research, he concluded that intonation is "virtually irrelevant as a question cue"...and that lexical-pragmatic indicators are more important for determining the question status of an utterance (2004: 313)

However, Gunlogson's case (2001) for rising declaratives is persuasive: the proposition of the rising declarative, also known as an echo-question ("you say that is a Volvo") "must already be known to the Addressee, and mutually known to be known....the context must provide some crucial information about how the Speaker expects his use of the declarative to be interpreted. Declaratives can be interpreted as questioning moves only if they *can't* be interpreted as telling"

(2001: 14; cf. Gunlogson 2003). The Alzheimer's speakers we reference here, and others in our larger corpus, are able to respond to declarative questions in general, and especially to *so*-prefaced declarative questions. Appendix A is an excerpt from a conversation in 2000 with Glory Mason, that shows her responses to a range of question and statement types.

So-prefaced statements are used in a variety of contexts: Waring (2002) notes that *so*-prefaced statements are likely to be interpreted as questions; in her study of seminar discussion, *so* is used for reformulation. Johnson (2003) finds them marking topic and organizing narrative in formal police interviews. We briefly examined two additional sources, each an online corpus: in the MICASE corpus (Michigan Corpus of Academic Spoken English), we sampled the genres of office hours, tutorials and interviews with at least one (female) speaker over the age of 50, and in the Oral History Interviews (OHP) and Charlotte Narrative and Conversation Collection (CNCC) in the New South Voices Collection (NSV), we sampled conversational interviews with men and women over 60. In the MICASE corpus, most of the *so*-prefaced statements were typically used to summarize and get the discussion back "on track" or in alignment with the agenda for at least one speaker, as in this exchange from the advising recordings:

MICASE Advising **Summary and move to get back "on track"**
[S1] that's uh there's something actually there's something like that in uh Bateson.
The one with the uh, the diagram.
S2: right. Okay.
S1: yeah. Okay well i think i think we're clicking on enough epistemological ideas here. Alright. so uh is this beginning to uh, to sound more like it might be a um, be or or become, a um, a thesis idea? [sic] (http://www.hti.umich.edu/m/micase/)

In the OHP and CNCC interviews by men of men and of women, and by women of men and of women, we noticed that *so*-prefaced statements were typically used by the interviewer for summary which could include inference either from what the interviewee had said (Marie Hicks, Clarissa Hampton) or from an inferred shared world knowledge (Henrietta Wallace):

Marie Hicks by Beth Sides (OHP)
MH: . . . And we met at that time.
BS: How old were you?

Clarissa Hampton by Kim Bailey (CNCC)
KB: Now did you go to Bellefonte Church then too?

MH: I was only about, I was about 14 at that first time.
BS: How old was he?
MH: And he was [pause] about 18. And later on my father's health improved and we went back to Kansas. And then it was a year after that that he came, my husband came, and we were married.
BS: *So you were what, 15, 16?*
MH: I was sixteen.
CH: Been going there all my life. I was born right there down there below the church.
KB: Uh huh.
CH: And I've been here all my life so I bought here and uh I guess I'll live here.
KB: Might as well. *So, you've never moved anywhere else.*
CH: No.
KB: Always lived here. (http://www.newsouthvoices.uncc.edu)

Henrietta Wallace by Ed Perzel (OHP)
HW: I graduated from Greensboro in 1931.
EP: *So you were in school during the Depression*?
HW: Yes, but also I stayed home a good many years with my family in Statesville

***So*-prefaced examples in conversations with Copeland and Mason**
The examples we present by Copeland illustrate her responses to *so*-prefaced statements offering resumption, inference, and single proposition available to be affirmed, with both a positive and a negative. In (2), she offers a *so*-prefaced statement as evaluation of what Bamberg would call a "small story": short, mundane, unnoticed at the time as a story, and incorporated into the discourse with the effect of positioning the participants relative to each other (see, for example, Bamberg 2004). In (3), she works with two different summary-inferences, the first signaling prior knowledge by "still," suggesting continuation, which she elaborates sufficiently in her response to scaffold an expansion of the topic and a second *so*-statement. The second *so*-statement contributes to Copeland's story line, so that she can offer an evaluative comment.

(1) **So + single choice resuming subject: affirms**
LM: So you're planning on a nice dinner with your friends tonight?
AC: Yes.
LM: And that's really good.
AC: Yeah. Very good.

(2) **So + negative single choice, the answer affirms with a negative**
FI: So you didn't stay there for very long?

AC: No, no. My husband was a general agent for an insurance company: ____. We were glad to move from there, because I thought it was an incentive to be lazy. My husband said that we couldn't prove that!

FI: Do you miss Boston?

AC: No. I really miss Winneka more than anywhere else. My husband was from ____, Massachusetts, which was a very old, old town: The statue of ____, one of his ancestors...

FI: I see.

AC: So, we touched a lot of places, but we liked it best of all

(3) So + inference from prior conversation; inference from current conversation

BD: So, are you still playing Canasta?

AC: Oh yes. Four nights a week.

BD: That's fantastic! You must be winning and having fun.

AC: Oh well. We don't play for money or anything.

BD: Well no, just for fun.

AC: Just for fun, yeah.

BD: So you have a regular group you play Canasta with?

AC: Different girls each night. Uh huh. There's lots to do here. There's no need of being bored.

Glory Mason's responses to *so*-prefaced statements, even as her impairments increased, remained similar. In (4) and (5), we offer *so*-prefaced summary of her information, which initiates a sequence of confirmation and reconfirmation leading to additional commentary, and perhaps supporting her evaluative closing to both "small stories." In (5), the questioner must expand the *so*-statement with a like-clause of comparison, for Mason to expand her comment. In (6), we see the *so*-statement "reminding" as it establishes an inference from information given a few minutes before.

(4) So + single choice summarizing information

BD: Wanted to show you a picture: last time we were here Gradie was talking about her dog – does this look like your dog?

GM: I can't even see it...to tell how it looks...like [mine?]> did you say?

LM: The dog...does the dog look like yours?

GM: I can't see what kind it is <*Unclear*> a poodle?

BD: Yeah.

GM: It's a teacup[?] poodle

BD: So it's a poodle.
GM: Mm-hmm.
LM: What was his name?
GM: Boogie.
LM: Boogie.
BD: Oh how nice.
GM: I don't know where w- where we got the name but that's what we called him, Boogie.

(5) So + single choice summarizing information, leading to expansion

G: See I told ___ when their daddy died, I said "you'll jus' have to take your daddy's place, I can't do it" an' he's – he's takes over like he – like he was my husband
B: So he takes over
G: Yeah
B: Like your husband
G: Don't hardly know how much your kids mean to you 'til your daddy – 'til their daddy's gone

(6) So + inference from current conversation

BD: Your dad, Edwards. You were Edwards before you married?
GM: Yeah
BD: So you lived in the area on a farm, the Edwards' farm
GM: uncl yeah I guess so reckon we liv – I reckon. I lived on a farm I – I don't know it's been a long time ago – I forgot

We summarize with this table:

What is presented to the listening partner	What is expected from the listening partner
So + single choice resuming subject	affirmation anticipated
So + single choice continuing subject	affirmation anticipated
So + negative single choice	affirmation with a negative is expected
So + single choice summarizing topic	affirmation anticipated
So + single choice presenting inference from summary of other speaker	affirmation anticipated
So + two choices, either summarized (given) or inferable	affirmation of first choice is expected

This discussion has examined selected discourse markers in the conversation of three women with Alzheimer's, noting the utility of indirect questions as conversation prompts keyed to summary and limited inference. As an Appendix, we include an excerpt from a conversational interaction with Glory Mason whose responses support the strength of both lexico-pragmatic cues and intonational contours. The interaction begins with formulaic routine and is replete with requests for clarification and repair, which weave direct and indirect questions together, finally scaffolding opinions and longer commentary from Mrs. Mason about the here-and-now – why, for example, she is living at Pleasant Meadows – while her cousin lives "down there" with her long-deceased parents, whom she mentions with a present tense that could be either the actual present or the historical present. After an exchange about her husband's preaching, she returns to her cousin for a there-and-then memory of creating playhouses under the old trees on the farm, pushing pine needles around to make the rooms. In the future, to ground similar conversations, training for professional caregivers might incorporate practice with indirect questions, particularly *so*-prefaced summaries, go-ahead signals or backchannels, and interactions which model and build on the more common functions of discourse markers. They may be little words; they don't get lost with time.

Appendix A Direct and indirect questions in a segment from 5-31-00: Glory Mason

BD: Good morning Glory Mason What's new? Good morning! How are you?

GM: I guess I'm all right. How about you?

BD: Doing fine! You've got new stuffed animals!

GM: Do what?

BD: These are new stuffed animals. Were you just talking about dogs in the room?

GM: Do what?

BD: Were you just talking about somebody having left dogs in the room?

GM: I don't know what I've been talking about! How do you expect me to remember what I've been talking about?

BD: I don't.

GM: Can you do that yourself?

BD: No. No I cannot.

GM: (*Laughs*) I haven't been talking much because I don't have anybody much to talk to, but the lady over there. She doesn't want to talk *all* of the time. And I don't either!

BD: Well, no. But you're doing ok?

GM: What?

BD: You're ok? You feel all right?

GM: Yes. I guess I do. You feel better when you're living at home and don't have to have somebody saying how and what to do!

BD: That's true!

GM: But people have gotten to breaking in down around Burnsville and all.

BD: Breaking in?

GM: My children got uneasy about me staying home by myself.

BD: So, after they got uneasy ....?

GM This is a good place to stay, as far as I know.

BD: But it's not home.

GM: No, it's not home. My Mother and Daddy are down there in their house. And my cousin, no.... I don't believe she's a cousin.... I forgot what kin she is, she's living in the house with them now. Her and her husband. I'm glad they are.

BD: You were telling me about your husband. Did he preach sermons?

GM: My husband?

BD: Would he be a preacher?

GM: Yes. He was a preacher that preached "hell hot and heaven beautiful!" (*They both laugh*)

BD: Heaven beautiful...

GM: Yes. "Hell hot and heaven beautiful!" That was one of his messages. I don't know... he preached all right. He was an Evangelistic-type preacher.

BD: I bet you went many places!

GM: Well, I had my family while I was young and couldn't go. I mean... you can't go with a bunch of little kids.

BD: No, you can't.

*** [*Noise in the hall followed by long pause*]

BD: "Hell hot and heaven beautiful!" What did he have to say about heaven?

GM: Well, what is there to say, except that hell is hot! Said it's a burning fire! And heaven's beautiful. Nothing else to say!

BD: That's true. And so he said it several times.

GM: You've got to believe that!

*** [*Noise in the hall followed by long pause*]

GM: I lived in Anson County before I moved up here.

BD: You like Anson County?

GM: Yes. That's where I was born and raised. Made playhouses and played.

BD: Playhouses?

GM: My Mother's sister lived there and she had girls about my age and we went over in the woods and made playhouses. We raked pine needles and made playhouses...it would rain all the way around. We would rake pine needles and then we would rake some more. Made beds and tables. (*Laughs*)....

References

Aijmer, Karin. (1987) "OH and AH in English Conversation." In *Corpus Linguistics and Beyond*, Willem Meijs (ed.), Amsterdam: Editions Rodopi, 61–86.

Aijmer, Karin. (2002) *English Discourse Particles: Evidence from a Corpus*. Amsterdam: John Benjamins.

Almor, A., Kempler, D., MacDonald, M., Andersen, E. & Tyler, L. (1999) "Why do Alzheimer patients have difficulty with pronouns? Working memory, semantics, and reference in comprehension and production in Alzheimer's disease." *Brain and Language* 67: 202–27.

Bamberg, Michael. (2004) "Talk, small stories, and adolescent identities". *Human Development* 47: 366–400.

Bayles, K. (2003) "Effects of working memory deficits on the communicative functioning of Alzheimer's dementia patients." *Journal of Communication Disorders* 36: 209–19.

Blakemore, Diana. (1996) "Are apposition markers discourse markers?" *Journal of Linguistics* 32: 325–47.

Byron, Donna & Peter A. Heeman. (1997) "Discourse Marker Use in Spoken Dialog." In *Proceedings of the 5th European Conference On Speech Communication and Technology*, Rhodes, Greece, September 1997, pp. 2223–6.

Clark, Herbert. (2002) "Speaking in time." *Speech Communication* 36: 5–13.

De Santi, Susan. (1993) "Formulaic Language in Aging and Alzheimer's Disease." PhD dissertation, City University of New York.

Ellis, D. (1996) "Coherence patterns in Alzheimer discourse." *Communication Research* 23: 472–95.

Erman, Britt. (2001) "Pragmatic markers revisited with a focus on you know in adult and adolescent talk." *Journal of Pragmatics* 33: 1337–59.

Fischer, K. & Drescher, M. (1996) "Methods for the description of discourse particles: contrastive analysis." *Language Sciences* 18: 853–61.

Fox Tree, Jean. (2002) "Interpretations of pauses and ums at turn exchanges." *Discourse Processes* 34(1): 37–55.

Fox Tree, Jean and Josef Schrock. (2002) "Basic meanings of you know and I mean." *Journal of Pragmatics* 34: 727–47.

Fuller, Janet. (2003) "Discourse marker use across three contexts: a comparison of native and non-native speaker performance." *Multilingua* 22: 185–208.

Fuller, Janet. (2003) "The influence of speaker roles on discourse marker use." *Journal of Pragmatics* 35: 23–45.

Grosz, Barbara J. & Candace L. Sidner. (1986) "Attention, intention, and the structure of discourse." *Computational Linguistics*, 12(3): 175–204.

Guendouzi, Jacqueline & Nicole Muller. (2001) "Intelligibility and rehearsed sequences in conversations with a DAT patient." *Clinical Linguistics and Phonetics* 15: 91–5.

Gunlogson, Christine. (2003) *True to Form: Rising and Falling Declaratives as Questions in English*. NY: Routledge.

Gunlogson, Christine. (2001) "Rising and falling declaratives." http://semanticsarchive.net 2001 07 29; last accessed June 19, 2004.

Hamilton, Heidi. (1994a) *Conversations with an Alzheimer's Patient.* Oxford: Oxford University Press.

Hamilton, Heidi. (1994b) "Requests for Clarification as Evidence of Pragmatic Comprehension Difficulty: The Case of Alzheimer's Disease." Pages 185–99 in Bloom, Ronald; Obler, Loraine; De Santi, Susan & Erlich, Jonathan (eds.), *Discourse Analysis and Applications: Studies in Adult Clinical Populations*. Hillsdale, NJ: Lawrence Erlbaum.

Heritage, John. (1984) "A Change-of-State Token and Aspects of Its Sequential Placement. In Dwight Atkinson & John Heritage (eds.), *Structures of Social Action*. Cambridge: Cambridge University Press, pp. 299–345.

Heritage, John. (1998a) "Oh-prefaced responses to inquiry." *Language in Society* 27: 291–334.

Heritage, John. (1998b) "Oh-Prefacing: A Method of Modifying Agreement/Disagreement." In Cecilia Ford *et al.*(eds.), *The Language of Turn and Sequence*. Oxford: Oxford University Press.

Johnson, Alison. (2003) "*So*. . .?: Pragmatic Implications of *So*-Prefaced Questions in Formal Police Interviews." In Janet Cotterill (ed.), *Language in the Legal Process*. Basingstoke and NY: Palgrave Macmillan, 91–110.

Jucker, Andreas. (1993) "The discourse marker well: A relevance-theoretical account." *Journal of Pragmatics* 19: 435–52.

Jucker, Andreas, Sara Smith & Tanja Luedge. (2003) "Interactive aspects of vagueness in conversation." *Journal of Pragmatics* 35: 1737–69.

Kasher, Asa (ed.). (1998) *Pragmatics: Critical Concepts, Vol. VI: Pragmatics, Grammar, Psychology, Sociology*. London: Routledge.

Katayama, H. (2001) "Beyond 'change-of-state': 'oh' as a facilitator of teacher–student interactions in an ESL conversation class." *Crossroads of Language, Interaction, and Culture* 4 [n.p.]. Retrieved originally from 2001 Conference Abstracts, http://www.sscnet.ucla.edu/clic/

Katzenberger, I. (2004) "The development of clause packaging in spoken and written texts." *Journal of Pragmatics* 36: 1921–48.

Keller, Joerg & Trixi Rech (1998) "Towards a modular description of the deficits in spontaneous speech in dementia." *Journal of Pragmatics* 29: 313–32.

Kempler, D. (1995) "Language Changes in Dementia of the Alzheimer Type." In R. Lubinsky (ed.), *Dementia and Communication: Research and Clinical Implications*. San Diego: Singular, pp. 98–114.

Kim, Jung Hee, Michael Glass, Reva Freedman & Martha Evens. (2000) "Learning the use of discourse markers in tutorial dialogue for an intelligent tutoring system." *Proceedings of the 22nd Annual Meeting of the Cognitive Science Society, CogSci, 2000*, Philadelphia, 262–7.

Linguistics Data Consortium. (2004) "Annotation guidelines: metadata (as resumptive of turn)," last accessed 25 June 2004 at http://www.ldc.upenn.edu/Projects/MDE/

Macaulay, Ron. (2002) "You know, it depends." *Journal of Pragmatics* 34: 749–67.

Mann, William C. & Sandra A. Thompson. (1988) "Rhetorical Structure Theory: towards a functional theory of text organization." *Text*, 8(3): 243–81.

Mahendra, N., Bayles, K.A. & Tomoeda, C.K. (1999) "Effect of an unfamiliar accent on the repetition ability of normal elders and individuals with Alzheimer's disease." *Journal of Medical Speech-Language Pathology*, 7: 223–30.

Müeller, N. (2003) "Intelligibility and negotiated meaning in interaction." *Clinical Linguistics and Phonetics*, 17: 317–24.

Norrick, Neal. (2000) *Conversational Narrative: Storytelling in Everyday Talk*. Amsterdam and Philadelphia: John Benjamins.

Obler, Loraine & De Santi, Susan. (2000) "Eliciting Language from Patients with Alzheimer's Disease." Pages 403–16 in Lise Menn & Nan Ratner (eds.), *Methods for Studying Language Production*. Mahwah, NJ: Lawrence Erlbaum.

Perkins, L., A. Whitworth & R. Lesser. (1998) "Conversing in dementia: a conversation analytic approach." *Journal of Neurolinguistics* 11: 33–53.

Perkins, Lisa, Anne Whitworth & Ruth Lesser. (1997) *Conversation Analysis Profile for People with Cognitive Impairment*. London: Whurr Publishers.

Perkins, Michael. (1998) "Is pragmatics epiphenomenal? Evidence from communication disorders." *Journal of Pragmatics* 29 (1998): 291–311.

Ramanathan, Vai. (1997) *Alzheimer Discourse: Some Sociolinguistic Dimensions*. Mahwah, NJ: Lawrence Erlbaum.

Rhys, Catrin & Nicola Schmidt-Renfree. (2000) "Facework, social politeness and the Alzheimer's patient." *Clinical Linguistics & Phonetics* 14: 533–43.

Ripich, D. & B. Terrell. (1988) "Patterns of discourse cohesion and coherence in Alzheimer's disease." *Journal of Speech and Hearing Disorders* 53: 8–15.

Ripich, D.N., Carpenter, B. & Ziol, E. (2000) "Conversational cohesion in men and women with Alzheimer's disease: a longitudinal study." *International Journal of Language and Communication Disorders*, 35(1): 49–65.

Rosenberg, Sheldon & Leonard Abbeduto. (1987) "Indicators of linguistic competence in the peer group conversational behavior of mildly retarded adults." *Applied Psycholinguistics* 8: 19–32.

Ryan, Ellen, S. Meredith, M. MacLean & J. B. Orange. (1995) "Changing the way we talk with elders: promoting health using the Communication Enhancement Model." *International Journal of Aging and Human Development* 41: 89–107.

Rendle-Short, Johanna. (2003) "So what does this show us?": analysis of the discourse marker 'so' in seminar talk." *Australian Review of Applied Linguistics* 26: 46–62.

Sabat, S.R. & Collins, M. (1999) "Intact social, cognitive ability, and selfood: A case-study of Alzheimer's disease." *American Journal of Alzheimer's Disease* 14: 11–19.

Safarova Marie, Swerts Marc. (2004) "On recognition of declarative questions in English." *Proceedings of the Speech Prosody 2004 Conference, Nara (Japan), March 23–26, 2004*, pp. 313–16.

Schiffrin, Deborah. (1987) *Discourse Markers*. Cambridge: Cambridge University Press.

Schourup, Lawrence. (2001) "Rethinking well." *Journal of Pragmatics* 33: 1025–60.

Schrock, J.C. & Fox Tree, J.E. (2000) " 'So' and 'and' in spontaneous speech." Poster presented at the 2000 meeting of the Psychonomic Society, New Orleans,

LA. Last accessed 15 June 2004 at http://psychology.gatech.edu/renglelab/chadpage/So_and_And.ppt
Small, J.A., G. Gutman, S. Makela & B. Hillhouse. (2003) "Strategies used by caregivers of persons with Alzheimer's disease during activities of daily living." *Journal of Speech, Language and Hearing Research*, 46: 353–67.
Stemmer, Brigitte & Paul Schoenle. (2000) "Neuropragmatics in the 21st Century." *Brain and Language* 71 (special issue).
Tappen, R.M., C. Williams-Burgess, J. Edelstein, T. Touhy & S. Fishman. (1997) "Communicating with individuals with Alzheimer's disease: an examination of recommended strategies." *Archives of Psychiatric Nursing*, 11(5): 249–56.
Temple, Valerie; Sabat, Steven; Koger, Rolf. (1999) "Intact use of politeness in the discourse of Alzheimer's sufferers." *Language and Communication* 19: 163–80.
Van de Craen, Piet. (2000) "Non-standard Dutch 'Allez' as a discourse particle." Conference on Discourse particles, modal and focal particles, and all that stuff, Universitaire Stichting, Brussels, December 2000. Abstract: http://odur.let.rug.nl/ ~vdwouden/particles/prog02.htm
Waring, Hansun Zhang. (2002) "Displaying substantive recipiency in seminar discussion." *Research on Language and Social Interaction*, 2002, Vol. 35: 453–79.
Wray, Alison. (2002) *Formulaic Language and the Lexicon*. Cambridge: Cambridge University Press.
Wray, A. & Perkins, M. (2000) "The functions of formulaic language: an integrated model." *Language and Communication* 20(1): 1–28.

8

Bad Times and Good Times: Lexical Variation over Time in Robbie Walters' Speech

Margaret Maclagan and Peyton Mason

R: Ohh I have my good days an' ba– (I uh) actually physically (I'm I have no–) I don't have any problems or anything
B: That is good
R: Hmm?
B: That's good
R: Mm-hmm
B: What's new?
R: Not– not anything really I'm tryin'... I have some bad times and good times I think

21 June 2000

In June 2000, Robbie Walters was talking to B who had visited him regularly for the last six months. She used one of her usual conversational gambits. Mr. Walters' reply summarizes what we discovered when we analyzed his speech:

B: What's new?
R: Not– not anything really I'm tryin'... I have some bad times and good times I think.

In this chapter we present the results of analyses of Mr. Walters' conversations over time with different interviewers. In order to examine possible changes over time, we carried out three different language analyses, conversational cohesion (following analyses used by Ripich *et al.* 2000; and Ripich & Terrell 1988) and lexical richness and word class analysis (following the methods used by Bucks *et al.* 2000). We comment also on differences between the results from interviews with two of the interviewers.

Background

There is general agreement that the language of patients with Alzheimer's disease deteriorates over time (see, e.g., Bayles & Kim 2003; Hopper, 2003; Kempler 1991). There is less agreement about the types of deterioration that can be expected. From early in the course of the disease, conversational partners report that Alzheimer's patients produce "empty" speech (Bayles *et al.* 1987; Kempler 1991). Syntactic and phonological abilities are relatively spared (Bucks *et al.* 2000; Kempler 1991; Ulatowska *et al.* 1988) while lexical, pragmatic and semantic abilities are affected from early stages of the disease (Bucks *et al.* 2000; Kempler 1991). Results of analyses of cohesion vary. Some researchers found that there were more cohesion errors (especially errors of reference) in Alzheimer's subjects than in normals (Ulatowska *et al.* 1988), while others found that cohesion deteriorated more slowly over time than did other language features (Ripich *et al.* 2000).

Researchers often use scripted linguistic tasks such as retelling stories or choose language tests to analyse the language of people with Alzheimer's disease (Bayles *et al.* 1987). Recently, there has been an emphasis on the need for analyses to be carried out on naturalistic conversations with speakers the subjects know well (Perkins *et al.* 1998; Temple *et al.* 1999; Ulatowska & Chapman 1991). In order to obtain more naturalistic language, researchers have used more or less structured interviews that attempted to duplicate spontaneous language. Bucks *et al.* (2000) used structured interviews with clear guidelines for the interviewers to follow. Ripich & Terrell (1988) used open-ended questions on three topics (family, daily activities and health) and Ripich *et al.* (2000) used a carefully structured interaction around a tea/coffee break during sessions of standardized testing in order to obtain relatively spontaneous language. Dijkstra *et al.* (2004) used five-minute conversational interviews with relatively unfamiliar interviewers who were instructed to use three conversational prompts (about the subject's family, life and day). However such situations may still underestimate the abilities of the speakers who are attempting to communicate in a relatively artificial situation, usually with someone they do not know well (see Temple *et al.* 1999). In addition, there is some evidence that the specific interviewer can affect the results obtained (see Perkins *et al.* 1998; Ramanathan-Abbot 1994).

The data used in this chapter come close to being fully spontaneous. They come from conversations between Mr. Walters and different researchers who were not known to Mr. Walters at the start of the project. However, because the visits were regular and initially frequent, the interviewers, especially B and G, soon became familiar figures to him. The data analysed here is therefore relatively naturalistic, and because both B and G became very familiar to Mr. Walters, we are also able to make some comparisons of Mr. Walters' language when talking to two different familiar interviewers.

Data

Robbie Walters, whose language is examined in this chapter and also in earlier chapters in the present volume, is a patient in the Alzheimer's Unit at Pleasant Meadows retirement living facility. He was recorded over a space of four years by several different interviewers. The data were collected as part of a larger project analysing speech from subjects with Alzheimer's disease. Four different interviewers interacted with Mr. Walters, usually alone, but sometimes two interviewers talked to him together. When two interviewers were involved in the conversation, the transcript was classified according to the interviewer who took the major part in the conversation. Neither of the authors of this chapter was an interviewer. Because this was a long-term project, the interviews occurred at irregular intervals. Initially, several interviews were conducted per month; however, as time progressed, and depending on interviewer availability, interviews were conducted less often. Additionally, there were several periods when interviews were missed due to Mr. Walters' physical health or because of restricted access to the nursing home from major outbreaks of the flu. The number of interviews therefore varies from year to year and from interviewer to interviewer. Sometimes there were several interviews in one month, but there are relatively few interviews in 2002 and 2003. This variability complicates both the analysis and the interpretation. Table 8.1 displays the transcripts used in this analysis. A total of 50 transcripts were included. They differed in length, depending to a large extent on how willing Mr. Walters was to talk and whether or not other residents were included in the conversation. They also varied in their degree of scripting. Those with B are virtually unscripted, as is the single interview with C. One interview with L contains an attempt to carry out a formal vocabulary assessment and many of the interviews with G contain routines involving counting in German and looking at pictures. The routines in G's interviews made it

Table 8.1 Interviews with Robbie Walters

Year	*Interviewer*				*Total*
	B	*G*	*L*	*C*	
2000	14	7	3	0	24
2001	5	13	2	0	20
2002	3	0	0	0	3
2003	1	1	0	1	3
Total	23	21	5	1	50

easier to obtain language, but reduced the spontaneity of the interactions. The interviews with B in particular come extremely close to spontaneous interactions.

Methodology

The interviews were transcribed and the transcriptions verified. Some interviews were coded for overlapping speech and features such as (*Laughter*) were noted. For the present analysis, the texts were divided into 24 different time periods of one month each. There were 12 time periods for 2000 (so that there were interviews in each month) seven for 2001, two for 2002 and three for 2003. The three interviews for 2003 each took place in a different month.

Three different types of analyses are reported here: cohesion analyses similar to those carried out by Ripich *et al.* (2000) and word class and lexical richness analyses following Bucks *et al.* (2000). The cohesion analysis requires hand coding and therefore is time consuming. Computers can automatically generate the data for the word class and lexical richness analyses. This makes them relatively easier to use for analysing larger texts.

For the cohesion analysis, pronoun reference, ellipsis and conjunction use were analysed. In each case a judgment was made as to whether or not the reference or ellipsis was appropriate and whether the reference or omitted material could be clearly understood from the context (see Halliday & Hasan 1976; Ripich *et al.* 2000). Reformulations, which occur in normal speech and are sometimes called *mazes* (Leadholm & Miller 1992), were not counted as ellipses (see Ripich *et al.* 2000 who separated ellipses from false starts). An analysis of noun initiation was also conducted to check whether Mr. Walters initiated noun use himself or whether he used nouns that had previously been used by other speakers. Nouns, pronouns, and conjunctions were also coded for appropriateness. Nouns were judged inappropriate if they had no relation to surrounding speech and pronouns were judged inappropriate if they did not relate appropriately to their referent (e.g., *she* instead of *they*). Responses from the interviewers were also used to help indicate whether or not the nouns and pronouns had been used appropriately. Further details of this analysis are given in Table 8.2 below. Full texts including interviewer speech as well as Mr. Walters' speech were used for the cohesion analysis. Interviews with B and G were included in the

Table 8.2 Areas included in the cohesion analysis

Analysis of nouns	
Noun initiation	whether Mr. Walters initiated noun use, or whether his nouns had been prompted by the interviewer
Appropriateness of noun use	whether or not the nouns used were appropriate in the context. Criteria for determining inappropriateness included interviewer response, or interviewer pause after Mr. Walters' utterance
Indefinite noun use	use of an indefinite noun such as *thing, product, lot, bit, stuff, area, kind, type, place, sort*, or *part*, where the context called for a definite one
Analysis of pronouns	
Type of pronoun reference	endophoric reference (where the reference is within the text) and exophoric reference (where the reference is outside the text) (Halliday & Hasan, 1976)
Clarity of pronoun use	whether or not the referent for the pronoun could be clearly retrieved
Appropriateness of pronoun use	whether or not the appropriate pronoun was used (including whether or not it matched the referent in person and number and was personal or indefinite as required by context)
Indefinite pronoun use	use of indefinite pronouns such as *one, anyone, anything*, or *someone*
Analysis of ellipsis	
Appropriateness of ellipsis	whether or not material elided could be easily recovered, as in *How are you? (I'm) Fine*
Analysis of conjunctions	
Use of conjunctions	use of coordinate and subordinate conjunctions, including the number of clauses that were joined in one utterance

cohesion analysis, because they provided the greatest amount of data and the best range of interviews over time. The single interview with C was also included in the cohesion analysis in order to provide more data for 2003. The first author carried out this coding with the assistance of a graduate research assistant who was studying speech pathology. Because the coders were not present when the original interviews took place, we were generous in our judgments of appropriateness. The results may therefore somewhat underestimate the lack of cohesion in the data.

In order to carry out the word class and lexical richness analyses, all interviewer speech, all non-linguistic indicators and all codes were removed, so that the texts contained only Mr. Walters' speech. Mr. Walters once knew some German but was never fluent in it. Because of this, G used routines in German in his interviews. Because of Mr. Walters' lack of fluency, any German words were also deleted from the texts. For the word class and lexical richness analyses, the texts were divided into 100 word units. For each unit, the number of items in the different word classes (nouns, verbs, pronouns and adjectives) were counted. A word class usage rate was obtained by dividing the word rate for each 100-word observation by each word class overall four-year mean. Brunet's Index, Honoré's Statistic, and Type–Token Ratio, to provide an indication of lexical variety and richness were calculated for the lexical richness analysis. The word class analyses and lexical richness measures (see Bucks *et al.* 2000) were all computer generated. The results of these analyses were subjected to statistical analyses to determine significant relationships between the items and to investigate change over time in Mr. Walters' language. All 50 texts were included in this analysis that was carried out by the second author.

The word class and lexical richness measures have been shown to discriminate well between participant groups (impaired versus non-impaired persons) when collection of language takes place at a single point in time (Bucks *et al.* 2000). We were concerned to investigate whether these measures would exhibit sensitivity to change when data collection took place over 48 months, with an individual person known to be impaired. We also wanted to compare these measures with results from the cohesion analyses. We asked the following questions:

a. Will features of cohesion change differently over time?
b. Will cohesion features change differently with different interviewers?
c. Will word classes (rate per 100 words of pronouns, nouns, adjectives and verbs) discriminate change longitudinally?

d. Will lexical richness discriminate change over time?
e. Will lexical richness be more sensitive a measure than word classes in monitoring change over time?

We expected that lexical richness as measured by word variety would deteriorate over time faster than word classes (Bucks *et al.* 2000) and that the cohesion analyses would be mixed. We expected ellipsis and conjunction use to deteriorate over time but other cohesion measures to remain constant in spite of being at a lower level than in normal speakers (Dijkstra *et al.* 2004; Mahendra & Arkin 2003; Ripich *et al.* 2000; Ulatowska *et al.* 1988).

Results: cohesion analysis

We consider first the results of the cohesion analysis. Table 8.2 indicates the four overall areas involved in this analysis. The data were initially analysed month by month. However, because of the variability in Mr. Walters' output, there was insufficient material in some cells for valid analysis. The data were then analysed year by year, and that data will be reported here. Each area will be considered in turn.

Results: analysis of nouns

The results of the noun initiation analysis are presented in Table 8.3. Noun initiation declined over time with one interviewer (B) but increased with the other (G). This seemed to reflect the different styles of the two interviewers (see discussion below). Chi-squared analysis

Table 8.3 Cohesion analysis. Percentage nouns initiated by Mr. Walters or produced after a prompt by the interviewer. Bold items indicate that the number is higher than would be expected

Year	*Interviewer B*				*Interviewer G*			
	Initiated	*Prompted*	*Total*	*n*	*Initiated*	*Prompted*	*Total*	*n*
2000	**73%**	27%	100%	547	59%	**41%**	100%	465
2001	**73%**	27%	100%	98	**71%**	29%	100%	521
2002	60%	**40%**	100%	68	0	0	0	0
2003	55%	**45%**	100%	20	**70%**	30%	100%	43

Note: Nouns were regarded as prompted if the interviewer used the same noun or a closely related noun in the preceding utterance.

showed that both patterns were significant; for B, $\chi^2=8.134$, df=2, p=0.043, for G, $\chi^2=16.78$, df=2, p=0.0002.

Appropriateness of noun use declined over time in interviews with G ($\chi^2=27.007$, df=2, $p<0.0001$) but showed no clear change in interviews with B, and there was no significant change in the relative number of indefinite nouns used over time with either interviewer.

Results: analysis of pronouns

There was no significant change in the use of endophoric and exophoric pronouns over time with either interviewer. Nor was there a significant change in the use of indefinite pronouns or in the clarity of pronoun reference over time for either interviewer. However, the use of the appropriate pronoun for the context decreased significantly over time for both interviewers, but the numbers involved were small. In order to clarify whether the appropriateness of Mr. Walters' noun and pronoun use declined over time, the analyses of appropriate noun and pronoun use were combined. The results are presented in Table 8.4. Inappropriate noun and pronoun use increased over time with both interviewers (for B, $\chi^2=20.59$, df=3, p=0.0001, and for G, $\chi^2=27.007$, df=2 and $p<0.0001$).

Results: analysis of ellipsis

Mr. Walters continued to use ellipsis throughout the time period studied. As shown in Table 8.5, there was no significant increase in the use of inappropriate ellipsis over this time ($\chi^2=4.286$, p=0.12). Substitution is a cohesive device very like ellipsis. In substitution, a proform is substituted for a word, phrase or clause to avoid repetition. Examples of substitution are *one* in utterances such as *John bought a cake and Mary*

Table 8.4 Cohesion analysis. Percent relevant and irrelevant nouns and pronouns used by Mr. Walters. Bold items indicate that the number is higher than would be expected

Year	*Interviewer B*				*Interviewer G*			
	Relevant	*Irrelevant*	*Total*	*n*	*Relevant*	*Irrelevant*	*Total*	*n*
2000	**95%**	5%	100%	1254	**94%**	6%	100%	924
2001	**98%**	2%	100%	209	90%	**10%**	100%	1149
2002	91%	**9%**	100%	137	0	0	0	0
2003	83%	**17%**	100%	40	82%	**18%**	100%	130

Table 8.5 Percent appropriate and inappropriate use of ellipsis over time. Bold items indicate that the number is higher than would be expected

	Appropriate ellipsis	*Inappropriate ellipsis*	*n*
2000	71%	**29%**	448
2001	**79%**	21%	68
2002 and 2003	**81%**	19%	69

Table 8.6 Percentage appropriate and inappropriate conjunction use over time

	Appropriate use	*Inappropriate use*	*n*
2000	80%	20%	284
2001	93%	7%	198
2002 and 2003	88%	12%	26

bought one too and *did* in *Mary had eggs for breakfast and John did too.* Although a detailed analysis was not made of substitution because it is relatively rare (Halliday & Hasan 1976), we noticed that Mr. Walters continued to use substitution appropriately throughout the time period analysed. Substitution could be a relatively easy process for someone with DAT to use, because it involves using an empty word rather than one with content. Nevertheless, Mr. Walters' use of substitution seemed to be appropriate. In particular, he did not use it excessively.

Analysis of conjunctions

Although conjunction use decreased over time (see Table 8.6), significant trends could not be identified. In particular, Mr. Walters continued to use both coordinate and subordinate conjunctions over the time period analysed, and he continued to use both types of conjunctions to join clauses. Relative use of subordinating conjunctions increased rather than decreased over the last two years. *And* joining phrases rather than clauses became more frequent over time.

Lexical richness and word class analysis

The linguistic measures identified by Bucks *et al.* (2000) were designed to analyze conversational, spontaneous speech. These measures have been shown to discriminate well between participant groups (impaired versus non-impaired persons) when collection of language takes place at a single point in time. We asked whether these measures would continue to exhibit sensitivity to change when collection took place over 48 months for one impaired person. The four measures used for word classes were noun rate, pronoun rate, verb rate, and adjective rate, and the three measures for lexical richness, which are often used in literary attribution studies (see Holmes 1998; Singh 2001), were Type–Token Ratio, Brunet's Index, and Honoré's Statistic.

Word classes

The usage index is a measure of variation that makes the changes relative to Mr. Walters. The index is constructed by taking the overall mean (all four years) and dividing it into the mean for each of the time periods (100-word units). The results show by percentage the extent that Mr. Walters changes on a period-to-period basis. Thus, 1.0 represents the mean usage rate. The dotted lines above and below the average line in Figures 8.1–8.8 mark one standard deviation of word usage from the mean. The R^2 measures the slope of the mean over the four years of interviews. For all of the measures the slope is flat, showing virtually no diminishing of word class usage rates.

The nouns (Figure 8.1), however, do exhibit a directional downward trend of diminished rate of usage for 2003. While this shift appears to be fairly certain, the sparseness of the data at this point makes it difficult to determine whether or not it represents a permanent decline in Mr. Walters' noun usage.

Adjectives (Figure 8.2) and verbs (Figure 8.3), unlike the nouns, maintain a consistent usage rate over the four-year period and particularly

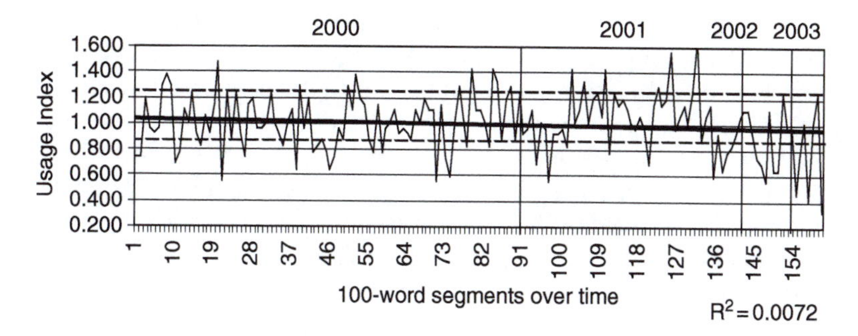

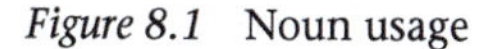
Figure 8.1 Noun usage

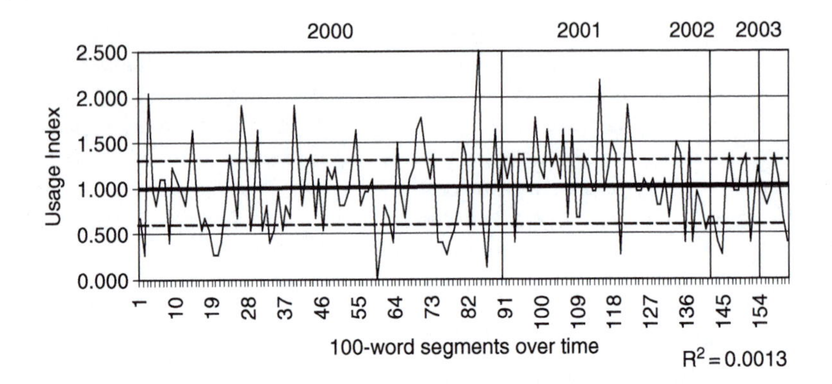

Figure 8.2 Adjective usage

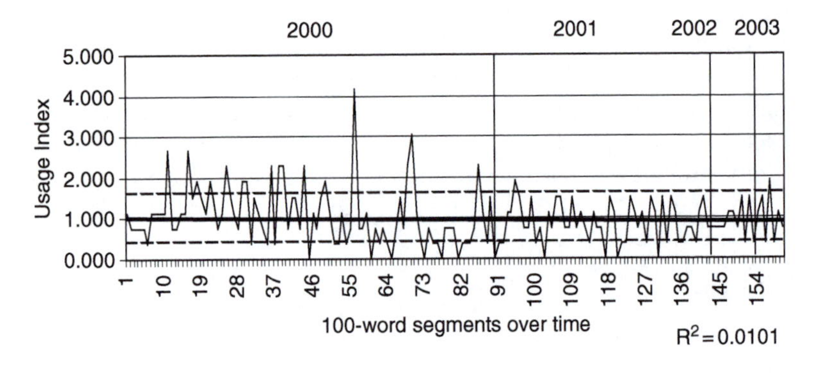

Figure 8.3 Verb usage

in the last one. The following figures illustrate that Mr. Walters' condition has greater swings, both positive and negative, in word usage on a month-to-month basis than from one year to the next one.

Mr. Walters' pronoun usage (see Figures 8.4–7) stayed intact over the four-year time frame. As with the nouns, adjectives and verbs, it displays fairly wide swings between interviews. Thus, the point in time one chooses to measure his capabilities may be more meaningful for determining his condition than, for instance, an infrequent periodic assessment.

These results show that Mr. Walters' ability to verbally communicate is more likely to manifest itself on a daily basis rather than as a

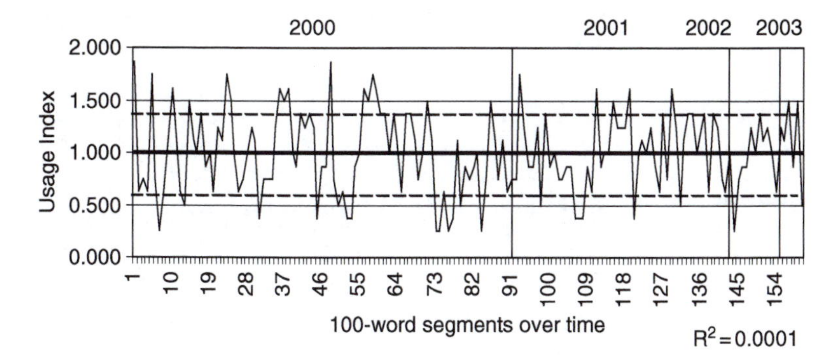

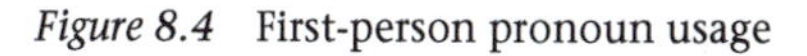
Figure 8.4 First-person pronoun usage

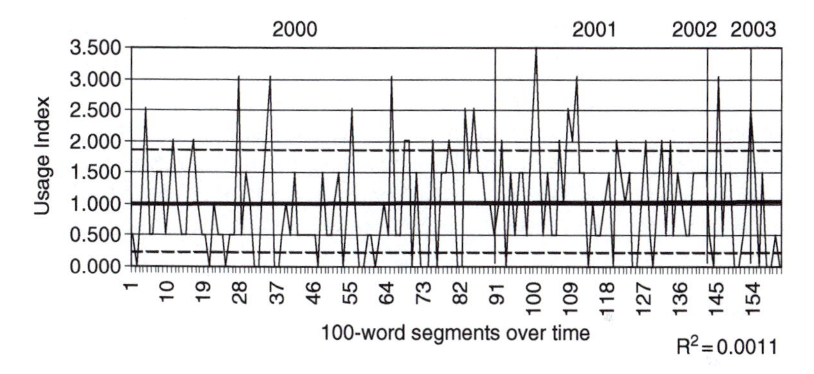

Figure 8.5 Second-person pronoun usage

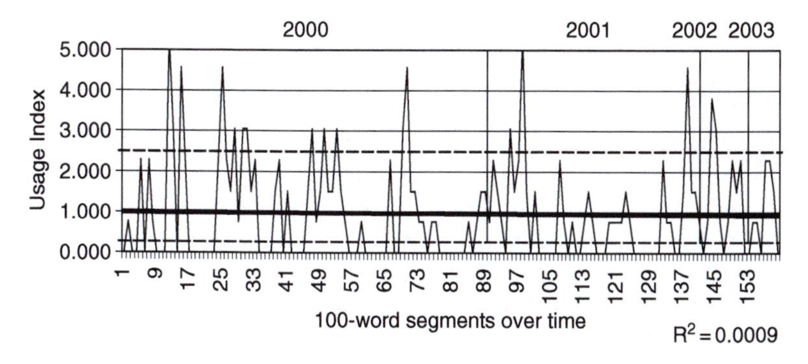

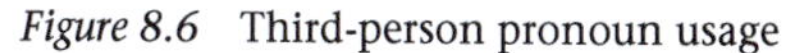
Figure 8.6 Third-person pronoun usage

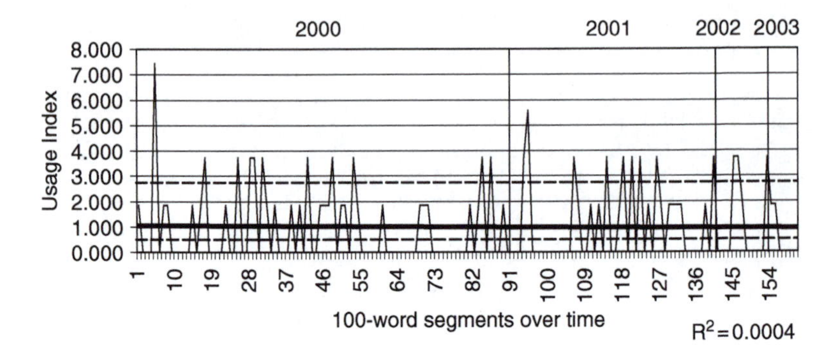

Figure 8.7 Indefinite pronoun usage

downhill slide through time. In short, communicatively he has good days and bad ones. His condition reflects a continuous ebb and flow of speech skills. While we cannot be certain if this pattern will persist, it appears to be indicative of having reached a possible plateau. The single possible exception may be his noun usage. This skill seems to be diminishing over the last year of his interviews. As we will see in the next section, Mr. Walters' Type–Token Ratio also exhibits a similar decreasing trend with respect to lexical richness. However neither of these changes is statistically significant.

Lexical richness

Three measures for lexical richness were used in this study: T–TR (Type–Token Ratio), Brunet's index, and Honoré's Statistic. Lexical richness is of special interest because of the emphasis in current research on progressive loss in Alzheimer's disease of general language features including vocabulary. Mr. Walters' lexical richness did not diminish in a linear manner over the four-year period; instead it varied from time to time, though in 2003 there appears to be the start of a downward movement.

Additionally, Mr. Walters' degree of lexical richness was consistent between the three measures. As is to be expected, there is almost unity between the three measures due to the 100-word standardization of Mr. Walters' transcripts (see Table 8.7).

Since the results of the three measures are nearly identical, we selected T–TR to simplify and illustrate Mr. Walters' lexical richness (see Figure 8.8). His T–TR displayed a similar pattern of vocabulary usage to the word class measures.

Table 8.7 Pearson's r correlations between lexical richness measures

	Type-Token Ratio	*Brunet's Index*
Brunet's Index	.96	–
Honoré's statistic	.97	.90

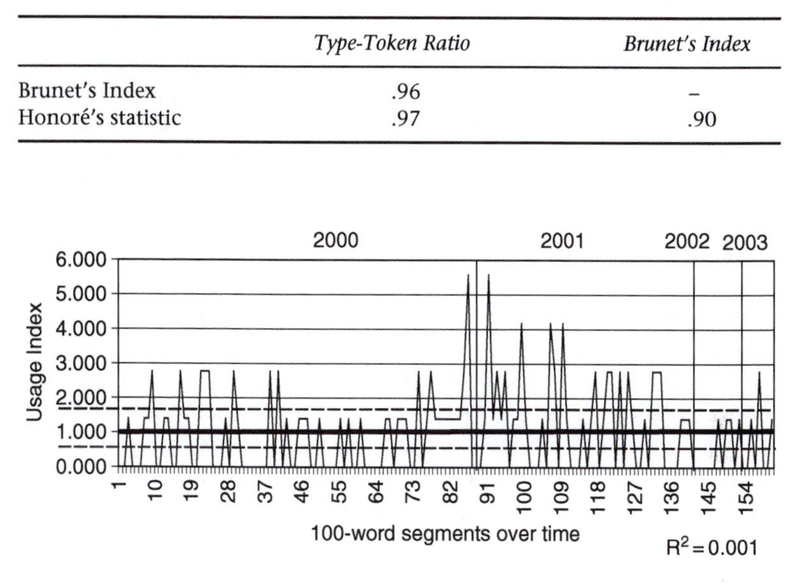

Figure 8.8 Interrogative pronoun usage

Discussion

The overwhelming result from these analyses is the variability of Mr. Walters' speech output over time. He does indeed have bad times and good times for speech, as well as for life in general. At the start of the analysis, we asked five questions:

a. Will features of cohesion change differently over time?
b. Will cohesion features change differently with different interviewers?
c. Will word classes (rate per 100 words of pronouns, nouns, adjectives and verbs) discriminate change longitudinally?
d. Will lexical richness discriminate change over time?
e. Will lexical richness be more sensitive a measure than word classes in monitoring change over time?

We expected that word classes and lexical richness would decline over time, with lexical richness showing a greater decline, but that many features of cohesion would not. We deal with each of these questions in turn.

Will features of cohesion change differently over time?

The cohesion features studied did show different patterns over time, but not the ones expected. While the number of conjunctions used declined over time, inappropriate conjunction use and inappropriate ellipsis use did not increase significantly (cf. Ripich *et al.* 2000). Nor did the use of indefinite nouns and pronouns increase significantly (see Mahendra & Arkin, 2003 for a similar result for indefinite nouns), even though greater use of indefinite words differentiates speakers with Alzheimer's disease from the normal elderly (see Dijkstra *et al.* 2004; Ulatowska *et al.* 1988). The significant changes within the cohesion analysis occurred in the number of noun initiations (see Table 8.3 above) and in the appropriateness of noun and pronoun use over time (see Table 8.4 above). Ripich *et al.* found that pronoun reference did not decrease significantly over time, although there was a downward trend (2000: 55–6). In the present study, the appropriate use of pronouns did decrease over time, and when noun and pronoun use was combined together, the decrease was significant for both interviewers.

Will cohesion features change differently with different interviewers?

Differences between the two interviewers can be seen in the different results for Mr. Walters' noun use over time. In the interviews with B, the number of nouns produced by Mr. Walters without a previous prompt decreased significantly over time. He continued to use nouns that had been used by B or were closely related to nouns B had used in immediately preceding conversational turns. However the appropriateness of his noun use with B did not decrease over time (see Table 8.3 above). G regularly used a set of pictures as part of his routines for encouraging Mr. Walters to remember the German words he once knew. Mr. Walters often named (in English) items in the pictures without previous models from G. This allowed his number of noun initiations to increase significantly over time in the interviews with G. When other interactional prompts ceased to be effective, the pictures still elicited some nouns. However, the appropriateness of Mr. Walters' noun use decreased over time with G. He consistently used more nouns and fewer pronouns per hundred words with B than with G, but the differences were not statistically significant. The other cohesion measures did not show significant differences between the two interviewers. Although both interviewers used pictures in their interactions with Mr. Walters,

G used them much more regularly than B. Because of this difference in interaction technique, it is difficult to separate out the effects of interviewer and technique on his language.

Will word classes (rate per 100 words of pronouns, nouns, adjectives and verbs) discriminate change longitudinally?

We expected that word class usage would decline less dramatically than lexical richness, but that it would decline. The trend measures for all the word classes analyzed were flat, indicating that there was little change over time. However the word class analysis showed marked swings from one session to the next. The noun analysis did show a downward trend in the last year (2003), but the small amount of data available in that year makes it hard to know if this is a genuine downward movement, or part of the overall variability of Mr. Walters' language.

Will lexical richness discriminate change over time?

The expected clear deterioration over time did not occur in the lexical richness analyses, with the three lexical richness analyses (T–TR, Brunet's Index and Honoré's Statistic) all showing similar patterns. Like the word class analyses, the lexical richness analyses emphasized the variability of Mr. Walters' language over time. Like the noun analysis, they showed slight downward trends in 2003, but again these trends are hard to interpret.

Will lexical richness be more sensitive a measure than word classes in monitoring change over time?

The analyses of Mr. Walters' interviews suggest that neither set of measures was more sensitive to his situation than the other. Lexical richness and word classes both exhibited period-to-period peaks and valleys of usage while over time there was little or no measurable diminishing of Mr. Walters' capabilities. For Mr. Walters, this seems to indicate that during the research period his condition did not worsen, at least as monitored by these measures. On a daily basis, however, his situation showed considerable swings up and down. Nevertheless, while his conversation capability did not diminish, his ability to converse on new subjects did. It appears that variants of the same topics were discussed multiple times.

In the analyses we have carried out, it has been difficult to find significant linguistic changes in Mr. Walters' speech, in terms of cohesive devices, word class usage or lexical richness. Only rate of noun usage, noun initiation and appropriates of noun and pronoun use showed any clear decline. In spite of the variability between conversations, it seemed as though he had reached a plateau and remained there (cf. Mahendra & Arkin 2003: 416 who comment on the similar slow decline over time of their subjects). However, in spite of these analyses, Mr. Walters' language did deteriorate over the time analyzed. As noted earlier, conversational partners complain of the emptiness of the speech of people with Alzheimer's disease (Bayles *et al.* 1987; Kempler 1991). Even though the areas we analyzed remained relatively constant, we saw an increase in emptiness in Mr. Walters' speech over time. The following three extracts all refer to his brother Wayne. The first extract from a conversation early in 2000 with B, demonstrates an abrupt shift in topic, but does contain new information not prompted by the interviewer. Similarly the second extract from 2001, also with B, has relevant content. In both of these extracts, Mr. Walters shows accurate time orientation; he knows he is talking about the past. The third extract, from the final conversation in the series with C late in 2003, has very little content and he talks about his brother in the present tense. He can still produce familiar nouns like *Wayne*, *home*, *mother*, and *dad* without a model from the interviewer, but he no longer volunteers any extra information. It is possible that analyses of topic such as information units (Dijkstra *et al.* 2004), topic comments (Mahendra & Arkin 2003) or propositions (Ulatowska *et al.* 1988) could have picked up these changes. In the extracts, three dots (...) indicate a pause, comments are in square brackets, and mazes are in parentheses.

Extract 1: 2000

B: What were some of the things you all grew?

R: Oh, I'd say I won't (I won't) always start with onions [*Laughs*] that's not, so that's just... of course she [his mother to whom he'd referred earlier] had radishes and (we had) we had a good bit from the garden early.

B: Radishes came early. (Your) Your weather in West Virginia was probably like mine in Kentucky.

R: Well Wayne always worked at the... oh... most of it was at the stores and things like that.

B: Mmhmm.

R: Cause he worked for Schultz's store and so they all stayed busy with the exception of the girls, didn't get in on a whole lot of anything.

Extract 2: 2001

B: I never heard you complain about anything.

R: Well, I don't think I've had anything, really worth, you know, something really creating a problem for me in life. I mean, I had a pretty happy home life and no ... just pretty content with (what) how things were going for me.

B: Well, you were born lucky.

R: Of course I went in the service early and Wayne didn't.

B: Yeah, you did go in the service early.

Extract 3: 2003

C: And let's see, you told me about your brother ...

R: Wayne?

C: Wayne! Yes ...

R: Yeah ...

C: You told me about Wayne.

R: Yes ...

C: Uh huh ...

R: Yeah, he's still home ...

C: He's home?

R: Yes.

C: And he is ... uh ...

R: With mother and dad.

C: Now was it just the two of you, you and Wayne? Did you have any other brothers or sisters?

R: Yeah, I have uh I have two sisters.

We had hoped that the use of almost completely spontaneous conversations would allow us clearly to indicate the deterioration in Mr Walters' speech over time. In spite of some significant changes over time, the measures used did not pick out the expected clear deterioration in Mr. Walters' speech. However all analyses showed extreme variation from time period to time period, and from interview to interview within time periods. This highlights the difficulties involved in obtaining valid analyses of the language of subjects with Alzheimer's disease, and demonstrates that analysing spontaneous speech, *per se*, does not guarantee accurate results.

The variation found between transcripts taken relatively close together in time indicates that results obtained are very dependent on the state of the subject at the time of the interviews. Neither extremely regular spontaneous samples nor periodic standardized testing would guarantee an accurate assessment of Mr. Walters' language abilities. At the simplest level, Mr. Walters' output in words varied from a high of 1391 words in September 2000 with G to a low of four words in January 2002 with B. However his output did not simply decline in quantity. In the last interview in the series, in August 2003, with a different interviewer, C, Mr. Walters still produced 387 words.

To give a concrete example of the difficulty in interpretation of these results, the apparent decline in noun usage found over the last year of analysis may represent a genuine decline in Mr. Walters' language, or it may be that, by chance, the samples were taken on increasingly "bad" days, where there had been "good" days in between. Mr. Walters' higher scores in the very last interview in the series, which was also the first interview with C, points to another difficulty in accurate on-going assessment of Alzheimer's patients. In this interview, C covered the sorts of topics that B and G covered in their first interviews three years earlier, asking about what Mr. Walters did during his earlier life. He still succeeded in providing some answers, using nouns that did not need to be prompted. This illustrates the plaintive comment from the daughter of another Alzheimer's patient: "It's hard to keep having the same conversation over and over again". For C, it was a new conversation, for B and G, it would have been going over ground that they had traversed many times before.

Conclusion

The results of this case study raise as many questions as they answer. Dijkstra *et al.* (2004: 275) indicate that the "discourse deficits that have been found in different types of discourse samples of adults with dementia…occur in interview style conversations as well". The variability that is apparent from one interview to the next with Mr. Walters indicates that single "interview style conversations" may be no better than formal assessments in obtaining an accurate picture of the language skills of a patient with Alzheimer's disease. Other researchers have noted the variability both among subjects with Alzheimer's disease and among normal, healthy elderly. Bayles *et al.* (1987) and Ryan (1991), for example, emphasize the variability among healthy older adults. Ulatowska and Chapman acknowledge the difference in performance of the two subjects they describe with Alzheimer's disease, and note that it "may be indicative of the lack of homogeneity among DAT [dementia of the Alzheimer's type] subjects in general" (1991: 124). Mahendra & Arkin (2003: 407) note that the scores for their four participants showed marked fluctuations from year to year. Bayles *et al.* (1987: 119–21) summarize the results of the Tucson long-term study of people with Alzheimer's disease. They indicate that the composite linguistic score (which did not include spontaneous conversation) discriminated between patients with Alzheimer's disease and healthy older adults, but that there was a great deal of variation in the scores of the patients with Alzheimer's disease, with some improving over time while others deteriorated. The current study carries the idea of variability farther and throws into stark relief the day-to-day variability within the language of one person. Like Ramanathan-Abbot (1994), this case study found differences in the language produced with two different interviewers. As in Bucks *et al.* (2000), there was some decline in lexical richness over time, and as in Ripich *et al.* (2000) cohesion changed slowly over time, but some areas did change. But the overriding impression at the end of the analysis is variability. And this variability indicates that all results from patients with Alzheimer's disease must be interpreted with great caution. Does an observed decrease in performance indicate a genuine decline in abilities, or were the different observations taken on days of greatly different productions within a radically fluctuating output? And, similarly, does a pair of equally good results indicate a levelling off of the effects of the disease, and a plateau, or are they too the result of a happy coincidence of analysis dates? The results obtained for patients with Alzheimer's disease over time may depend more on the chance occurrence of "bad times and good times" than on the discriminating ability of the analyses performed or the naturalness of the conversation obtained.

Acknowledgement

We would like to thank the original interviewers for making the transcripts of the interviews available to us, Irfon Jones for computing help and Hannah Chacko for assistance with the cohesion analysis.

References

Bayles, K.A., Kaszniak, A.W., & Tomoeda, C.K. (1987) *Communication and Cognition in Normal Aging and Dementia.* London: Taylor & Francis.

Bayles, K.A. & Kim, E.S. (2003) "Improving functioning of individuals with Alzheimer's disease: emergence of behavioural interventions." *Journal of Communication Disorders*, 36: 327–43.

Bucks, R.S., Singh, S., Cuerden, J.M., & Wilcock, G.K. (2000) "Analysis of spontaneous, conversational speech in dementia of Alzheimer type: evaluation of an objective technique for analysing lexical performance." *Aphasiology*, 14: 71–9.

Dijkstra, K., Bourgeois, M.S., Allen, R.S., & Burgio, L.D. (2004) "Conversational coherence: discourse analysis of older adults with and without dementia." *Journal of Neurolinguistics*, 17: 263–83.

Halliday, M.A.K. & Hasan, R. (1976) *Cohesion in English*. London & NY: Longman.

Holmes, D.I. (1998). "The evolution of stylometry in humanities scholarship." *Literary and Linguistic Computing*, 13(3): 111–17.

Hopper, T.L. (2003) " 'They're just going to get worse anyway': perspectives on rehabilitation for nursing home residents with dementia." *Journal of Communication Disorders*, 36: 345–59.

Kempler, D. (1991) "Language Changes in Dementia of the Alzheimer Type." In R. Lubrinski (ed.), *Dementia and Communication* (pp. 99–114). Philadelphia: B.C. Decker.

Leadholm, B.J. & Miller, J.F. (1992) *Language Sample Analysis: The Wisconsin Guide*. Madison, Wisconsin: Wisconsin Department of Public Instruction.

Mahendra, N. & Arkin, S. (2003) "Effects of four years of exercise, language, and social interventions on Alzheimer discourse." *Journal of Communication Disorders*, 36: 395–422.

Perkins, L., Whitworth, A., & Lesser, R. (1998) "Conversing in dementia." *Journal of Neurolinguistics*, 11(1–2): 33–53.

Ramanathan-Abbot, V. (1994) "Interactional differences in Alzheimer's discourse: an examination of AD speech across two audiences." *Language in Society*, 23: 31–58.

Ripich, D.N., Carpenter, B., & Ziol, E.W. (2000) "Conversational cohesion patterns in men and women with Alzheimer's disease: a longitudinal study." *International Journal of Language and Communication Disorders*, 35(1): 49–64.

Ripich, D.N. & Terrell, B.Y. (1988) "Patterns of discourse cohesion and coherence in Alzheimer's disease." *Journal of Speech and Hearing Disorders*, 53: 8–15.

Ryan, E.B. (1991) "Normal Aging and Language." In R. Lubrinski (ed.), *Dementia and Communication* (pp. 84–97). Philadelphia: B.C. Decker.

Singh, S. (2001) "A pilot study on gender differences in conversational speech on lexical richness measures." *Literary and Linguistic Computing*, 16(3): 251–64.

Temple, V., Sabat, S., & Kroger, R. (1999) "Intact use of politeness in the discourse of Alzheimer's sufferers." *Language and Communication*, 19: 163–80.

Ulatowska, H.K., Allard, L., Donnell, A., Bristow, J., Haynes, S.M., Flower, A., *et al.* (1988) "Discourse Performance in Subjects with Dementia of the Alzheimer Type." In H.A. Whitaker (ed.), *Neuropsychological Studies of Nonfocal Brain Damage* (pp. 108–31). NY: Springer-Verlag.

Ulatowska, H.K. & Chapman, S.B. (1991) "Discourse Studies." In R. Lubrinski (ed.), *Dementia and Communication* (pp. 115–32). Philadelphia: B.C. Decker.

Part II

Text and Context

9

Communication Enhancement for Family Caregivers of Individuals with Alzheimer's Disease

Kerry Byrne and J.B. Orange

Family caregivers of individuals with dementia have been studied extensively over the past two decades by health and psychosocial care researchers representing a diverse range of professions and working in a variety of rehabilitation, long-term care and community settings. Speech-language pathologists, occupational therapists, communication specialists, nurses, social workers and psychologists, among other professionals, have all contributed to the extant literature regarding the experiences and needs of family caregivers of individuals with dementia. Given the expanding prevalence of dementia in the aging population worldwide, health and psychosocial care professionals are increasingly faced with the need to provide treatment to individuals with some form of dementia. These professionals are necessarily working in close proximity with family caregivers in community contexts because individuals with dementia continue to live at home and are usually cared for by a family member, typically a spouse (Canadian Study of Health and Aging Working Group 1994).

This chapter begins by emphasizing the differences between family caregivers of individuals with dementia and family caregivers of individuals without dementia. Next, a summary is provided of communication-related difficulties identified by family caregivers of individuals with dementia. Following this, current published communication enhancement education and training interventions for family caregivers of individuals with Alzheimer's disease (AD) are discussed. The chapter concludes with a brief discussion of the importance of developing and testing communication enhancement education and training programs for family caregivers, and highlights directions for future research.

Family caregiver comparisons

Comparisons between family caregivers of individuals with dementia and those caring for individuals without dementia reveal that the former group spend more time in their roles, have more tasks associated with caregiving and have typically been providing care for a longer period of time (Ory *et al.* 1993). Family caregivers of individuals with dementia, particularly those who are spouses, also must complete a range of unique, complex and diverse caregiving tasks while often simultaneously having to address their own age-related multiple health conditions and psychosocial needs. The implications of the differences between the two caregiver groups are far reaching in that family caregivers of individuals with dementia suffer more negative health and emotional consequences than do caregivers of individuals without dementia. For instance, Clipp & George (1993) compared family caregivers of individuals with dementia versus family caregivers of patients with cancer. They found that even after controlling for baseline differences in age, duration of illness and patient symptoms, the dementia caregivers scored lower on measures of physical and emotional health, reported greater use of psychotropic drugs and had a more compromised social life. Similarly, Leiononen *et al.* (2001) reported that spouses of individuals with dementia who had noncognitive psychiatric symptoms reported higher burden scores than those spouses caring for relatives who were depressed.

Perhaps one of the most comprehensive comparisons to date of family caregivers of individuals with dementia versus family caregivers of individuals without dementia was conducted by Ory *et al.* (1999). The authors found from their nationally representative survey that caregivers of individuals with dementia were more likely to report employment complications such as having to take on less demanding roles at work, having to take early retirement, turning down a promotion, losing job benefits, and, in some cases, having to give up work entirely. A greater proportion of the family caregivers of individuals with dementia stated they had to give up pleasurable personal activities, had less time for other family members, and suffered greater degrees of family conflict. These caregivers also scored higher on measures of emotional and physical strain and were more likely than the family caregivers of individuals without dementia to indicate that their mental and physical problems are the result of caregiving. Notably, even after Ory *et al.* controlled for sociodemographic variables and level of caregiving involvement, they found that being caregivers for individuals with dementia is still a significant predictor of both physical and emotional strain and financial hardship.

It is clear that family caregivers of individuals with dementia suffer a wide range of health, vocational and emotional consequences associated with their caregiving roles. When compared to caregivers who look after relatives with other health conditions, family caregivers of those with dementia still score considerably worse on many indicators of well-being. Nonetheless, family caregivers prefer to care for their loved ones with dementia at home rather than placing their relatives in long-term care settings (Canadian Study of Health and Aging Working Group 1994). Not surprisingly, researchers often conclude that services and interventions for family caregivers of individuals with dementia must be targeted specifically to meet their unique needs in order to optimize the care they provide and to minimize the negative consequences of the caregiving process.

Communication-related difficulties reported by family caregivers

Communication difficulties are among some of the first symptoms of dementia reported by caregivers (Bayles & Tomoeda 1991). Research suggests that these difficulties not only occur early and frequently, but that caregivers rate them as highly problematic (Orange 1995). The fact that family caregivers of individuals with dementia must cope with their relative's language and communication problems differentiates their experiences of caregiving from family members caring for relatives who do not have language and communication problems. In the following section we summarize the salient results from studies that addressed family caregivers' perceptions of the communication and language difficulties of their relatives with dementia.

Standardized and non-standardized language and communication assessments of individuals in the three clinical stages of dementia (i.e., early, middle and late) have produced a wealth of detailed findings. It is now well accepted that nearly all individuals with dementia exhibit language and communication difficulties that vary in accordance to its clinical stages (Bayles 2003; Bayles & Kaszniak 1987; Bayles *et al.* 1992). Few studies, however, have fully examined caregivers' perceptions of language and communication-related difficulties and the extent to which these difficulties impact family caregivers. One of the earliest studies to identify communication and language as problematic for family caregivers of individuals with dementia was conducted by Rabins *et al.* (1982). The authors asked caregivers the following question: "What is the biggest problem you have in caring for the patient?" Twenty-two different problems were identified with 68 percent of family caregivers reporting the occurrence of communication difficulties and 74 percent reporting these difficulties as a significant problem in caring for their relative with dementia. In addition, catastrophic reactions were among the highest rated as problematic (89 percent of caregivers). Many other researchers now suggest that catastrophic reactions may be precipitated by communication breakdowns that lead to frustration and agitation for individuals with dementia (Clarke 1991; Orange 1995).

Quayhagen & Quayhagen (1988) examined behaviors reported to be stressful by husbands, wives and daughters who provided care for a relative with dementia. While the different groups of family caregivers varied in their perceptions of the degree to which they found behaviors such as embarrassing acts (e.g., swearing or socially inappropriate comments) to be problematic, the groups did not differ on ratings of repetitive question asking. That is, husbands, wives and daughters all found repetitive questions to be stressful. In fact, the total percentages for all three groups resulted in repetitive questions being identified as the most frequently cited stressful behavior. Orange (1995) conducted structured interviews with family caregivers of individuals with AD. Most family caregivers identified communication difficulties in the semantic domain (i.e., word finding) as highly problematic and over half reported difficulty dealing with problems in discourse and semantics. Moreover, caregivers reported feeling frustration, loneliness, guilt, embarrassment and social isolation as a direct result of their relative's communication difficulties. One caregiver commented as follows:

> ...I feel lonely sometimes because it's not the same. There is nothing coming back in the way of conversation... but it's the talking I miss. I miss the conversation. I miss discussing the why's and wherefore's and trying to figure out why those people did that or what's gonna happen down here. I miss that a whole lot. (Orange 1995, p. 184)

Similar findings were reported by Williamson & Schulz (1993) in their investigation of family caregivers of individuals with dementia regarding the specific caregiving situations that were most difficult for them to face over the last month. The inability to communicate with their relative as they had done in the recent past was the second most commonly cited difficult situation (memory difficulties were identified as the most frequent difficulty). In a study aimed at determining difficulties related to breakdown of communication, Powell *et al.* (1995) asked family caregivers about the prevalence of 32 dementia-related language and communication symptoms. Many caregivers reported that their relatives asked the same question a number of times, had difficulty following a conversation when a group of people are talking, had trouble keeping a conversation going and told the same story or piece of information a number of times. In a national European study, Murray *et al.* (1999) conducted 280 semi-structured interviews with family caregivers of individuals with dementia. Spouses reported that the most frequent difficult aspect of their relative's dementia that they had to cope with was the loss of their relative's language and communication. Sixty-eight respondents, or 24 percent of the caregiver participants, identified language and communication problems as one of the most difficult aspects of dementia to deal with, superseding other aspects such as loss of memory, aggression and uncooperative/stubborn behavior.

Employing focus group methodology, Small, Geldhart & Gutman (2000) asked family caregivers of individuals with dementia to consider the communication difficulties they experience with their relatives during activities of daily living (ADL) and during other activities in which they find communication is the most problematic. Caregivers reported that conversation involving personal life issues and using the telephone are the most affected. Half of the caregivers stated that using the bathroom, planning an agenda, locating an item and meal preparation also are problematic. The majority of activities were mentioned either once or twice by each caregiver with the exception of conversation which was identified, on average, three times per caregiver. The following quote from one of the participants exemplifies how the lack of conversation can affect caregivers:

> Sometimes I find it difficult when he is home that he sits there not talking. I have to do the talking, but it's like to the wall. I don't get anything back (p. 297).

The results from studies reviewed above provide compelling evidence that family caregivers find the language and communication problems of their relatives with dementia to be a stressful component of their caregiving. Over a decade ago, Clarke (1991) reported that there were no published studies regarding whether, how much and when the communication changes associated with AD affect the emotional status of caregivers. While several studies have since been published on the language and communication difficulties reported by family caregivers of individuals with AD, the relationship between reported difficulties and outcome measures such as depression, life satisfaction, quality of life and other variables commonly assessed in caregiver literature (e.g. sex, education, family member status, etc.) has not been studied from an empirical perspective (Orange 1995; Savundranayagam *et al.* 2004).

More recently, Turner & Street (1999) pointed out that even though the literature is replete with information on caregiver burden and stress, very little is known about adequate ways to assess the mental health and psychosocial needs of caregivers of individuals with dementia. They conducted a study aimed at identifying potential needs based on family caregivers' perceptions. They found that in order to cope successfully with individuals with AD, caregivers want information on how best to communicate with their relative (60 percent) and how best to manage difficult behaviors (70 percent). The caregivers' requests for information on these topics support the calls made by researchers and clinicians over the past decade for the development, testing and implementation of communication enhancement education and training programs for caregivers of individuals with AD (Orange *et al.* 1995; Small & Gutman 2002). The caregivers' demands for this information is especially poignant considering the assertions made by various authors that difficult behaviors can be precipitated by communication breakdowns (Clarke, 1991; Santo Pietro & Ostuni 2003). Savundranayagam *et al.* (2005) recently established a link between difficult behaviors and communication problems. They found that communication problems mediated the relationship between care recipient cognitive and functional status and problem behaviors. Interestingly, Hart & Wells (1997) found that residents with dementia demonstrated more instances of agitation (i.e., general restlessness, strange noises, etc.) when the language used in caregivers' commands was more complex than the comprehension ability of the residents. Notably, problem behaviors also have been consistent predictors of various measures of burden for dementia caregivers (e.g., Clyburn *et al.* 2000; Croog *et al.* 2001), highlighting the importance of studying the relationship between communication difficulties and attempts to reduce challenging or dysfunctional behaviors.

Interventions for family caregivers of individuals with AD

Family caregivers of individuals with AD have been studied at great length regarding the objective and subjective experiences of their roles as well as the factors that influence these experiences such as age, sex and socioeconomic status (Schulz *et al.* 1995). Not surprisingly, intervention research aimed at assisting family caregivers has become a heavily studied area. The search to reduce caregivers' levels of stress and depression and to improve quality of life and well-being has been carried out typically in the form of psychoeducational interventions, support groups, respite/adult day care, psychotherapy initiatives, environmental modifications and interventions that target care-receiver behaviors (e.g., pharmalogical evaluations) (Sörenson *et al.* 2002; Schulz *et al.* 2002).

A recent series of cogent reviews has been published about the effects, strengths and limitations of caregiver interventions (Bourgeois *et al.* 1996; Dunkin & Anderson-Hanley 1998; Schulz *et al.* 2002), including a meta-analysis that examined the effectiveness of interventions with caregivers of older adults with and without dementia (Sörenson *et al.* 2002). Sörenson *et al.* demonstrated that caregiver interventions do indeed impact outcomes, albeit modestly in many cases with most effect-sizes ranging from small to moderate. The authors found that psychoeducational and psychotherapeutic interventions demonstrated the most consistent effects on caregiver burden, depression, uplifts of caregiving, subjective well-being, caregivers' ability/knowledge and effects on care recipients' symptoms. Psychoeducational programs consisted of structured sessions involving lectures and group discussions along with supportive written materials. A trained leader provided information about the disease process, local, regional and national resources and services, and effective caregiver response patterns to the problems that arise from the disease. Psychotherapeutic interventions, on the other hand, included mostly a cognitive-behavioral approach including therapists teaching caregivers to challenge negative thoughts and to develop problem-solving abilities. Stronger intervention effects in these studies were noted for changing caregivers' ability/knowledge than were found for burden, depression, uplifts and care receiver symptoms, indicating that some outcomes are more sensitive to change than others.

Overall, Schulz and colleagues (2002) report that interventions for family caregivers of individuals with dementia have resulted in outcomes such as delayed institutionalization and high value ratings of the interventions by caregivers. The common theme among the majority of the intervention studies included in the reviews outlined above is that they include a variety of effective techniques and approaches that successfully educate and train caregivers.

Importance of communication education and training interventions

Bourgeois (1998) highlighted the crucial importance of the American Speech-Language and Hearing Association's 1988 recommendation that speech-language pathologists need to provide assistance to families to facilitate understanding of communication in individuals with dementia. Tomoeda (2001), in a review of speech and language assessments for individuals with dementia, reiterated the need for care plans to include counseling and training to both healthcare and family caregivers regarding the impairments and retained abilities of individuals with dementia. Moreover, she emphasized the need to ensure that caregivers understand the functional communication abilities (i.e., everyday communication skills) of individuals with dementia.

Richter *et al.* (1995) noted that because expressive language skills remain relatively strong in the early stages of dementia, caregivers may have unrealistic expectations of their relative's communicative abilities, expecting them, for example, to be able to retrieve new and old information in conversations. Clarke (1991) reported that due to the fluctuations and moments of lucidity, caregivers may think the individual with dementia is "faking it". Similarly, Bourgeois (2002) stated that caregivers often can be in denial about the speech and language problems that their relatives with AD experience. She referred to an example of a prototypical caregiving spouse who reports that his wife is intentionally upsetting him by repeating questions. In this case, Bourgeois noted that the spouse is not aware of the communication problem associated with repetitive question asking but is reacting to the literal meaning of the question because it is intact grammatically and there are no overt difficulties with speech and language. Other authors also have commented on the need to ensure that caregivers have the requisite knowledge about communication-related changes and realistic expectations when communicating with relatives with AD (Clarke 1991; Tomoeda 2001).

It is essential that family caregivers understand the complexity of the communication changes associated with AD and its progression and the need for them to make changes in their own communication with their family member with AD (Clarke 1995). More specifically, family caregivers need to understand that *they* must be the ones to adapt their communication because their relatives with AD often are not able to do so on their own, particularly as the disease progresses. Richter *et al.* (1995) point out that caregivers can be taught communication-enhancing strategies and that strengthening existing skills could result in family members being able to provide care at home for a longer period of time, thus delaying institutionalization.

The importance of educating and training family caregivers of individuals with AD on language and communication issues is underscored by their reports that friends and family become the central communication partners to individuals with AD and that family caregivers go to great lengths to make accommodations they perceive to be useful in helping their relative communicate (Orange 1995). Caregivers reportedly employ various strategies even without formal communication training. For instance, Baxter *et al.* (2002) reported that wives whose elderly husbands had AD or a related dementia identified strategies such as emphasizing nonverbal communication, increasing interpretive work, using nursing home staff as information mediators, and limiting contact with the spouse. Richter *et al.* (1995) used a focus group methodology to identify verbal and nonverbal strategies used by both family members and nursing home staff to manage fearfulness, agitation and wandering. Strategies used by both types of caregivers included adjusting the environment (e.g., reducing background noise) and reassuring the individual with dementia both verbally and nonverbally.

In addition to caregivers' reports of the communication strategies that they use, several researchers observed caregivers interacting with individuals with AD and found that caregivers employ communication strategies similar to those described above. In a study examining the effectiveness of communication strategies used by caregivers, Small *et al.* (2003) audio-recorded interactions between caregivers and individuals with AD. Family caregivers frequently used strategies such as eliminating distractions and approaching slowly to their relative with AD. Jansson *et al.* (2001) made field observations of married couples in their homes to determine the activities performed by caregivers when caring for a spouse with AD or a related dementia. They identified four themes, one of which was communication. They then divided communication into the categories of communication-interaction, empathy-thoughtfulness, and nearness. For instance, the spouse caregivers answered questions empathetically and participated in interactions even though their relatives with dementia could no longer manage a continuous conversation. Interestingly, conversations were achieved despite the fact that the individuals with dementia repeated the same questions many times and that conversations addressed the same topic. The following observation of a couple exemplifies the capability of caregivers to employ various strategies to facilitate communication:

> Mr.B does a crossword in the newspaper. Sometimes he asks his wife about some words. Even if Mrs B does not know and can not be helpful any longer, it seems as if she feels she is participating and enjoys it (p. 808).

In an experimental setting, Kemper *et al.* (1994) found that spouses of individuals with probable AD spontaneously employ a specialized speech register when communicating information to their relatives during a picture description task. The spouses provided more semantic content, increased semantic redundancy and used a simpler syntax consisting of fewer embedded clauses when communicating with their relatives with AD. Kemper *et al.* noted that participants with AD were more likely to be successful at the task when these appropriate speech accommodations were used. Such accommodations may prevent communication breakdown or, at the very least, facilitate successful communicative interactions. Kemper *et al.*'s study highlights the self-initiated efforts made by caregivers and has important implications regarding the ability of caregivers to learn to adjust their speech and language to facilitate communication. However, further work is needed to determine the circumstances under which speech, language and communication accommodations support productive, meaningful and rewarding communication between family caregivers and their relatives with AD.

Despite the fact that family caregivers reported the use of supportive communication strategies and have been observed using them in natural and experimental settings, we know little about the ways in which they obtain information about communication and AD and the relative effectiveness of various strategies. Small and Gutman (2002) identified ten strategies commonly recommended in the literature and constructed a questionnaire in which caregivers were asked how often they used these strategies. Although caregivers reported using some of the published strategies, there were a number of recommended strategies that caregivers reported using infrequently such as eliminating distractions, and establishing and maintaining eye contact. Thus, while the studies reviewed above indicate that family caregivers seemingly spontaneously employ supportive communication strategies, the communication strategies commonly used by caregivers are not always the ones identified in the literature, suggesting that published research on effective communication strategies is not reaching family caregivers. Small and Gutman (2002) argued that due to discrepancies between published sources of communication strategies and caregiver reports of strategy use, there is a need for better education and training programs for caregivers on effective communication enhancement strategies. The following section outlines the few published communication-related interventions conducted with family members caring for relatives with AD.

Communication interventions for family caregivers

A review of specific recommended speech, language and communication strategies is beyond the scope of this chapter. However, Orange (2001) provides a detailed summary of language, cognitive, conversation, speech and nonverbal strategies, as well as considerations of the influence of communication environments, medication and emotions, interactive style, roles and relationships and perceptions and attitudes on communication for individuals with AD. A wide range of communication strategies is thought to be useful for communicating with individuals with AD. While some researchers have conducted investigations of their effectiveness, empirical evidence of their usefulness remains limited (Small *et al.* 2003; Small *et al.* 1997). Strategies believed to be the most helpful have been included in several published communication education and training programs for lay audiences (e.g., Ripich 1996; Santo Pietro & Ostuni 2003).

The majority of communication education and training interventions developed to date have been implemented with a range of formal health care providers (Bourgeois 1990; Bourgeois 1993; Bourgeois *et al.* 2003) and nursing aides (e.g., Hoerster *et al.* 2001; Burgio *et al.* 2001; McCallion *et al.* 1999). A review of these programs is beyond the scope of this chapter. The following is a discussion of interventions directed to family members providing care to relatives with AD.

Bourgeois and colleagues reported the results of three communication interventions involving family members and external memory aids such as memory wallets and conversation notebooks (Bourgeois 1992; Bourgeois *et al.* 2002; Bourgeois *et al.* 1997). Following in-depth interviews with family members, clinicians created memory wallets and notebooks, which included pictures and written information about the individual with AD. Bourgeois (1990; 1992) conducted interventions whereby family members trained their relatives with AD to use memory wallets in conversations. The purpose of the memory wallets was to increase the use of on-topic statements about personal factual information by individuals with AD in conversations with familiar partners. In both studies, participants with AD made more accurate factual statements as a result of using the memory wallets. Bourgeois commented that memory wallets have the potential to be used as a tool to facilitate communication because they focus on the retained communication abilities of individuals with AD. She also demonstrated that some individuals with AD used the wallets in conversations, even without any training.

In a related study, Bourgeois *et al.* (1997) conducted an intervention with family caregivers of individuals with AD that aimed to reduce repetitive verbalizations made by the participants with AD. Family caregivers were instructed to use consistently a cuing system; that is, to direct their relative's attention to a memory notebook or to have their relative read responses from prepared memo-boards or index cards when they repeated words or phrases. The number of times the individuals with AD repeated the same statement or question was reduced successfully as a direct result of the training.

More recently, Bourgeois *et al.* (2002) investigated the effects of two skills-based interventions. One group of family caregivers received training about how to modify coping behaviors (self-change group) and another group received training about how to modify the problem behaviors of individuals with AD (patient-change group). In the patient-change group, caregivers were trained to use written cuing systems for repetitive verbal behavior to manage this stressful behavior. The patient-change skills training reduced the frequency of problem behaviors both at initial testing time and follow-up. The authors concluded that caregivers were well able to learn and implement strategies that focus on changing patient behavior, providing further support so that family caregivers can learn and implement communication strategies.

Ripich and colleagues developed a communication intervention for family members using her FOCUSED program (Ripich, 1996). FOCUSED is an acronym for *Face* the person *Orient* to the topic, *Continue* the topic, *Unstick* communication blocks, *Structure* with questions, *Exchange* conversation, and use *Direct* statements; all words that represent communication strategies that can be used by caregivers to enhance interaction with their relatives with AD. Ripich, Ziol and Lee (1998) implemented the FOCUSED program with family caregivers of participants with AD to determine whether communication training had an impact on caregiver outcomes such as affect, depression, health, general hassles, and hassles related to communication. They also examined the maintenance of these changes over time. Family caregivers received eight hours of communication training over a four-week period including information about AD and communication, the correction of misconceptions surrounding communicating with their AD relative and learning communication strategies. As a result of the program, caregivers demonstrated a decrease in communication related hassles over time and increased their knowledge about AD and communication. However, communication training did not impact affect, depression, health status or general hassles.

Ripich, Kercher, Wykle, Sloan and Ziol (1998) conducted a comparable study using the FOCUSED program with African-American and Caucasian family caregivers. Again, following communication training, caregivers reported more knowledge of AD and also expressed increased satisfaction in their communication with their relatives. However, only the African-American caregivers had decreased levels of stress after the intervention. In a related study, Ripich *et al.* (1999) examined whether the FOCUSED program could impact the type of questions used by family caregivers during interactions with AD relatives. After training, there was a trend for the family members to use fewer open-ended questions, resulting in conversations that flowed more smoothly. The studies by Ripich and colleagues demonstrate that it is possible to educate and train family caregivers on language and communication issues in AD, although results may vary according to the intended outcomes.

Orange and Colton Hudson (1998) conducted a case study whereby a spousal caregiver of an individual with AD received a communication education and training program in her home twice weekly over a 12-week period. Immediately following completion of the intervention the spouse used more effective techniques to resolve communication problems, there were fewer instances of communication breakdown between the spouse and her husband, she held fewer negative emotional responses to challenging behaviors, and she had a positive response to the implementation and completion of the program. The authors pointed out that further study is needed to test empirically the hypothesis that communication-enhancement education and training programs can improve communication, reduce stress and burden and ultimately optimize the quality of life of caregivers and individuals with AD.

More recently, Done and Thomas (2001) evaluated two formats of a communication skills training program for family caregivers of individuals with AD. The first intervention consisted of two, one-hour communication training sessions led by a speech-language pathologist who presented video scenes depicting communication breakdowns between two people acting as a wife and a husband with dementia. A second training video with the same scenarios was then shown depicting successful communication encounters. The clinician explained and reinforced the strategies used in the videos. The second intervention included only a booklet outlining communication problems and strategies similar to those presented in the first intervention. The ability to manage communication problems at home increased significantly for both groups of family caregivers, but only participants in the face-to-face first intervention reported greater improvement in knowledge about effective communication strategies. Both groups reported that learning the various strategies made caring easier and had affected their relationship with their relative with AD. Qualitative analyses of participants' comments to the training revealed overt positive benefits of the communication training and showed that caregivers could have prevented various behaviors had they known the strategies earlier. For example, one caregiver commented as follows:

> ...if I could have seen that video with the right and the wrong way it would have saved me no end of problems (p. 819)

It is clear that the results of studies on communication enhancement education and training for family caregivers of individuals with AD discussed above provide preliminary evidence of their usefulness. The range of methodologies and targeted strategies, in concert with the position that education (i.e., knowledge acquisition) and training (i.e., knowledge translation, transfer and implementation) are key elements in interventions, underscore the crucial importance of incorporating a dynamic and comprehensive approach to communication enhancement for family caregivers of individuals with AD. What is lacking among most of the studies, however, is a unified and scientifically sound theoretical base upon which to build and shape new communication behaviors of the family caregivers. The Communication Enhancement Model of Aging, first proposed by Ryan *et al.* (1995) and applied to individuals with AD (Orange *et al.* 1995), is a theoretical framework that emphasizes individualized strategies that are conceptualized within and shaped by multiple contextual influences. The use of this model is warranted relative to future empirically derived studies designed to examine the effectiveness of communication enhancement education and training programs for family caregivers of individuals with AD.

Conclusions and future implications

It is well accepted that communication difficulties are a major problem for family members caring for individuals with AD, that family caregivers can learn to adapt their communication, and that communication education and training programs have the potential to alleviate communication breakdown, increase communicative encounters and decrease negative outcomes for caregivers. However, relative to the magnitude of contentions that communication education and training corresponds to positive results for family caregivers, few researchers have examined the direct effects of these interventions on outcomes deemed to be important in caregiver literature. For example, although researchers have hypothesized that communication training may decrease stress levels and delay institutionalization (Gallagher-Thompson *et al.* 1997; Richter *et al.* 1995) further research is needed to demonstrate these types of relationships. In order to make the role of communication education and training for family caregivers relevant in administrative and policy domains, it will be necessary to demonstrate empirically that communication and education training does indeed impact positively on the caregiving role.

Researchers developing and implementing communication education and training programs are in a key position to take advantage of the many recommendations made by those who have conducted interventions with family caregivers of individuals with AD. Bourgeois *et al.* (1996) provided a comprehensive review of a number of interventions for caregivers of individuals with AD concluding that researchers and clinicians must consider controlling for a variety of factors in order to improve the quality of future interventions. For instance, caregiver characteristics, such as the familial relationship, sex and age must be accounted for, given the differential impact of interventions on these groups. Treatments need to be described in full detail to allow for replication and must be monitored closely to ensure the homogeneity of delivery to caregivers.

Bourgeois *et al.* (1996) emphasize the necessity of choosing appropriate outcome measures to demonstrate treatment efficacy and ensuring that the outcome measures reflect the treatment goals. For these reasons, it is important to consider the outcome measures that have been used to date to assess the effectiveness of communication interventions. Findings by Savundranayagam *et al.* (2005) suggested that communication problems mediated the relationship between care-recipient status and problem behaviors. If as asserted, the relationship between communication problems and problem behaviors can be demonstrated and elaborated, the most revealing outcome measures need to be used to confirm and explore this relationship.

Furthermore, measuring global outcomes in family caregiver communication enhancement programs, such as quality of life, may not accurately reflect the influence of the interventions given large individual differences among dementia caregivers (Dunkin & Anderson-Hanley 1998) and extensive differences regarding problems they find stressful (Rabins *et al.* 1982). Schlosser (2004) evaluated the use of Goal Attainment Scaling (GAS) as a means of evaluating change in the treatment of communication disorders. GAS is an individualized, criterion-referenced measure that allows for the evaluation of individual progress towards goals. It was developed in the 1960s (Kiresuk & Sherman 1968) and has since been used by a variety of professions, including geriatric rehabilitation (Stolee *et al.* 1992; Stolee *et al.* 1999) where it has demonstrated strong psychometric properties (Stolee *et al.* 1992; Stolee *et al.* 1999; Stolee *et al.* 2001) and clinical utility (Stolee *et al.* 1999). Schlosser (2004) concluded that, despite its limitations, GAS is a viable option for evaluating individualized change and that it can be used with a variety of populations and interventions. The process of GAS would allow family caregivers to identify the unique and specific communication difficulties they are experiencing, and help them to achieve individualized goals that would reflect positively on the communicative relationship they have with their relatives with AD. Nonetheless, assessments of the language and communication of participants must be comprehensive, individualized and include standardized and non-standardized approaches that also include functional communication measures (Lubinski & Orange 2000; Orange & Kertesz 2000). Moreover, the interventions, as noted above, must be motivated by and based on a sound theoretical model of communication and aging. Unfortunately, few studies to date have achieved such integrated harmony.

The assurance of an individualized approach could be attained by the use of single case experimental designs (Hayes 1981), for instance, tailoring a communication education and training program to a single caregiver and taking repeated measurements over time to assess the effectiveness of the intervention on various communication-related outcomes. Bourgeois (1998) also suggested that single subject designs could be used as a first step to developing effective communication education and training programs. Determining the components of the programs that are useful for family caregivers could potentially inform interventions aimed at larger groups of caregivers.

In addition to choosing appropriate outcome measures and considering alternative methodologies, types of significance also are an issue in caregiver intervention research (Schulz *et al.* 2002). Statistically significant differences between treatment and control groups do not always suggest clinically meaningful results (see Ogles *et al.* 2001 for a review of clinical significance). Using a threshold for clinical significance allows researchers to examine the "real-world" meaning of change pre- and post-intervention. Schulz and colleagues (2002) incorporated various measures of clinical significance because of their importance to ensure meaningful results for family caregivers. They established broad criteria and multiple approaches to assess clinical significance including the evaluation of symptomatology (measures of depression or anxiety), quality of life (burden, stress, social support, etc.), social significance (e.g., service utilization, time spent on caregiving tasks) and social validity (caregiver intervention evaluation ratings). Moreover, they evaluated the clinical meaning of effect size measures used in the literature. They found evidence of clinically significant outcomes in dementia intervention research, including social validity, improvements in depressive symptoms and delayed institutionalization.

The importance of developing and testing empirically theoretically sound, comprehensive and individually tailored communication enhancement education and training programs for family caregivers of individuals with AD cannot be overemphasized. Researchers and clinicians alike are professionally bound and morally obligated to create and offer widely communication education and training programs that achieve meaningful changes in the day-to-day activities and lives of family caregivers to individuals with AD.

References

Baxter, L.A., Braithwaite, D.O. Golish, T.D. & Olson, L.N. (2002) "Contradictions of interaction for wives of elderly husbands with adult dementia." *Journal of Applied Communication Research*, 30(1): 1–26.

Bayles, K.A. (2003) "Effects of working memory deficit on the communicative functioning of Alzheimer's dementia patients." *Journal of Communication Disorders*, 36: 209–19.

Bayles, K. & Kaszniak, A. (1987) *Communication and Cognition in Normal Aging and Dementia*. Boston: College Hill/Little, Brown.

Bayles, K.A. & Tomoeda, C.K. (1991) "Caregiver report of prevalence and appearance order of linguistic symptoms in Alzheimer's patients." *The Gerontologist*, 31: 210–16.

Bayles, K., Tomoeda, C., & Trösset, W. (1992) "Relation of linguistic communication abilities of Alzheimer's patients to stage of disease." *Brain and Language*, 42: 454–72.

Bourgeois, M. (1990) "Enhancing conversation skills in Alzheimer's disease using a prosthetic memory aid." *Journal of Applied Behavior Analysis*, 23: 29–42.

Bourgeois, M. (1992) "Evaluating memory wallets in conversations with patients with dementia." *Journal of Speech and Hearing Research*, 35: 1344–57.

Bourgeois, M. (1993) "Effects of memory aids on the dyadic conversation of individuals with dementia." *Journal of Applied Behavior Analysis*, 26: 77–87.

Bourgeois, M.S. (1998) "Functional outcome assessment of adults with dementia." *Seminars in Speech and Language*, 19(3): 261–79.

Bourgeois, M.S. (2002) "Where is my wife and when am I going home? The challenge of communicating with persons with dementia." *Alzheimer's Care Quarterly*, 3(2): 132–43.

Bourgeois, M.S., Camp, C.J., Rose, M., White, B., Malone, M., Carr, J., & Rovine, M. (2003) "A comparison of training strategies to enhance use of external aids by persons with dementia." *Journal of Communication Disorders*, 36: 361–78.

Bourgeois, M.S., Burgio, L.D., Schulz, R., Beach, S., & Palmer, B. (1997) "Modifying repetitive verbalizations of community-dwelling patients with AD." *The Gerontologist*, 37: 30–9.

Bourgeois, M.S., Schulz, R., Burgio, L.D. & Beach, S. (2002) "Skills training for spouses of patients with Alzheimer's disease: Outcomes of an intervention study." *Journal of Clinical Geropsychology*, 8(1): 53–73.

Bourgeois, M.S., Schulz, R., & Burgio, L. (1996) "Interventions for caregivers of patients with Alzheimer's disease: a review and analysis of content, process, and outcomes." *International Journal of Aging and Human Development* 43: 35–92.

Burgio, L., Allen-Burge, R., Roth, D., Bourgeois, M., Dijkstra, K., Gerstle, J., Jackson, E., & Bankester, L. (2001) "Come talk with me: Improving communication between nursing assistants and nursing home residents during care routines." *The Gerontologist*, 41: 449–60.

Canadian Study of Health and Aging Working Group. (1994) "Patterns of caring for people with dementia in Canada." *Canadian Journal of Aging*, 13: 470–87.

Clarke, L.W. (1991) "Caregiver Stress and Communication Management in Alzheimer's Disease." In D. Ripich (ed). *Handbook of Geriatric Communication Disorders* (pp. 127–42). Austin, Texas: Pro-Ed.

Clarke, L.W. (1995) "Interventions for persons with Alzheimer's disease: Strategies for maintaining and enhancing communicative success." *Topics in Language Disorders*, 15(2): 47–66.

Clipp, E.C. & George, L.K. (1993) "Dementia and cancer: A comparison of spouse caregivers." *The Gerontologist*, 33: 534–41.

Clyburn, L.D., Stones, M.J., Hadjistavropoulos, T., & Tuokko, H. (2000) "Predicting caregiver burden and depression in Alzheimer's disease." *Journal of Gerontology: Social Sciences*, 55B(1): S2–S13.

Croog, S.H., Sudilovsky, A., Burleson, J.A., & Baume, R.M. (2001) "Vulnerability of husband and wife caregivers of Alzheimer disease patients to caregiving stressors." *Alzheimer Disease and Associated Disorders*, 15(4): 201–10.

Done, D.J. & Thomas, J.A. (2001) "Training in communication skills for informal carers of people suffering from dementia: a cluster randomized clinical trial comparing a therapist led workshop and a booklet." *International Journal of Geriatric Psychiatry*, 16: 816–21.

Dunkin, J.J. & Anderson-Hanley, C. (1998) "Dementia caregiver burden: a review of the literature and guidelines for assessment and intervention." *Neurology*, 51(Suppl 1): S53–S60.

Gallagher-Thompson, D., Dal Canto, P.G., Darnley, S., Basilio, L.A., Whelan, L., & Jacob, T. (1997) "A feasibility study of videotaping to assess the relationship between distress in Alzheimer's disease caregivers and their interaction style." *Aging and Mental Health*, 1: 346–55.

Hart, B.D. & Wells, D.L. (1997) "The effects of language used by caregivers on agitation in residents with dementia." *Clinical Nurse Specialist*, 11(1): 20–3.

Hayes, S.C. (1981) "Single case experimental design and empirical clinical practice." *Journal of Consulting and Clinical Psychology*, 49(2): 193–211.

Hoerster, L., Hickey, E., & Bourgeois, M. (2001) "Effects of memory aids on conversations between nursing home residents with dementia and nursing assistants." *Neuropsychological Rehabilitation*, 11: 399–427.

Jansson, W., Nordberg, G., & Grafström, M. (2001) "Patterns of elderly spousal caregiving in dementia care: an observational study." *Journal of Advanced Nursing*, 34(6): 804–12.

Kiresuk T.J. & Sherman, R.E. (1968) "Goal Attainment Scaling: A general method for evaluating comprehensive community mental health programs." *Community Mental Health Journal*, 4: 443–53.

Kemper, S., Anagnopoulos, C., Lyons, K., & Heberlein, W. (1994) "Speech accommodations to dementia." *Journal of Gerontology: Psychological Sciences*, 49(5): P223–P229.

Leinonen, E., Korpisammal, L., Pulkkinen, L., & Pukuri, T. (2001) "The comparison of burden between caregiving spouses of depressive and demented patients." *International Journal of Geriatric Psychiatry*, 16: 387–93.

Lubinski, R. & Orange, J.B (2000) "A Framework for the Assessment and Treatment of Functional Communication in Dementia." In L.E. Worrall & C.M. Frattali (eds.) *Neurogenic Communication Disorders: A Functional Approach* (pp. 220–46). NY: Thieme.

McCallion, P., Toseland, R., Lacey, D., & Banks, S. (1999) "Educating nursing assistants to communicate more effectively with nursing home residents with dementia." *The Gerontologist*, 39(5): 546–58.

Murray, J., Schneider, J., Banerjee, S., & Mann, A. (1999) "Eurocare: a cross-national study of co-resident spouse carers for people with Alzheimer's disease: II – a qualitative analysis of the experience of caregiving." *International Journal of Geriatric Psychiatry*, 14: 662–7.

Ogles, B.M., Lunnen, K.M., & Bonesteel, K. (2001) "Clinical significance: History, application, and current practice." *Clinical Psychology Review*, 21(3): 421–46.

Orange, J.B. (1995) "Perspectives of Family Members Regarding Communication Changes". In R. Lubinski (ed.). *Dementia and Communication* (pp. 168–86). San Diego, CA: Singular Publishing.

Orange, J.B. (2001) "Family Caregivers, Communication and Alzheimer's Disease." In M.L. Hummert & J.F. Nussbaum (eds.). *Aging, Communication and Health* (pp. 224–48). New Jersey: Lawrence Erlbaum.

Orange, J.B. & Colton-Hudson, A. (1998) "Enhancing communication in dementia of the Alzheimer's type: caregiver education and training." *Topics in Geriatric Rehabilitation*, 14(2): 56–75.

Orange, J.B. & Kertesz, A. (2000) "Discourse analyses and dementia". *Brain and Language*, 71: 172–4.

Orange, J.B., Ryan, E.B., Meredith, S.D., & MacLean, M.J. (1995) "Application of the communication enhancement model for long-term care residents with Alzheimer's disease." *Topics in Language Disorders*, 15: 20–35.

Ory, M.G., Hoffman, R.R., Yee, J.L., Tennstedt, S., & Schulz, R. (1999) "Prevalence and impact of caregiving: a detailed comparison between dementia and nondementia caregivers." *The Gerontologist*, 39: 177–85.

Powell, J.A., Hale, M.A., & Bayer, A.J. (1995) "Symptoms of communication breakdown in dementia: carers' perceptions." *European Journal of Disorders of Communication*, 30: 65–75.

Quayhagen, M. & Quayhagen, M. (1988) "Alzheimer's stress: Coping with the caregiving role." *The Gerontologist*, 28: 391–6.

Rabins, P., Mace, N., & Lucas, M. (1982) "The impact of dementia on the family." *Journal of the American Medical Association*, 24(8): 333–5.

Richter, J.M., Roberto, K.A., & Bottenberg, D.J. (1995) "Communicating with persons with Alzheimer's disease: experiences of family and formal caregivers." *Archives of Psychiatric Nursing*, 9: 279–85.

Ripich, D.N., Ziol, E., Fritsch, T., & Durand, E.J. (1999) "Training Alzheimer's disease caregivers for successful communication." *Clinical Gerontologist*, 21(1): 37–56.

Ripich, D.N., Ziol, E., & Lee, M.M. (1998) "Longitudinal effects of communication training on caregivers of persons with Alzheimer's disease." *Clinical Gerontologist*, 19: 37–55.

Ripich, D.N., Kercher, K., Wykle, M., Sloan, D.M., & Ziol, E. (1998) "Effects of communication training on African-American and white caregivers of persons with Alzheimer's disease." *Journal of Aging and Ethnicity*, 1: 163–78.

Ripich, D.N. (1996) *Alzheimer's Disease Communication Guide: The FOCUSED Program for Caregivers*. San Antonio, TX: The Psychological Corporation.

Ryan, E.B., Meredith, S.D., Maclean, M.J., & Orange, J.B. (1995) "Changing the way we talk with elders: promoting health using the communication enhancement model." *International Journal of Aging and Human Development*, 41(2): 89–107.

Santro Pietro, M.J. & Ostuni, E. (2003) *Successful Communication with Persons with Alzheimer's Disease: An In-service Training Manual*, 2nd edition. St. Louis, MO: Butterworth-Heinemann.

Savundranayagam, M.Y., Hummert, M.L., & Montgomery, R.J.V. (2005) "Investigating the effects of communication problems on caregiver burden." *Journal of Gerontology: Social Sciences*, 60 B(1), 548–55.

Schlosser, R.W. (2004) "Goal attainment scaling as a clinical measurement technique in communication disorders: a critical review." *Journal of Communication Disorders*, 37: 217–39.

Schulz, R., O'Brien, A., Czaja, S., Ory, M., Norris, R.N., Martire, L.M., Belle, S.H., Burgio, L., Gitlin, L., Coon, D., Burns, R. Gallagher-Thompson, D., & Stevens, A. (2002) "Dementia caregiver intervention research: in search of clinical significance." *The Gerontologist*, 42(5): 589–602.

Schulz, R. O'Brien, A.T., Bookwala, J., & Fleissner, K. (1995) "Psychiatric and physical morbidity effects of dementia caregiving: prevalence, correlates and causes." *The Gerontologist*, 35: 771–91.

Small, J.A., Geldart, K., & Gutman, G. (2000) "Communication between individuals with dementia and their caregivers during activities of daily living." *American Journal of Alzheimer's Disease and Other Dementias*, 15: 291–302.

Small, J.A. & Gutman, G. (2002) "Recommended and reported use of communication strategies in Alzheimer caregiving." *Alzheimer Disease and Associated Disorders*, 16(4): 270–78.

Small, J.A., Gutman, G., Makela, S., & Hillhouse, B. (2003) "Effectiveness of communication strategies used by caregivers of persons with Alzheimer's disease during activities of daily living." *Journal of Speech. Language, and Hearing Research*, 46: 353–67.

Small, J.A., Kemper, S., & Lyons, K. (1997) "Sentence comprehension in Alzheimer's disease: effects of grammatical complexity, speech rate, and repetition." *Psychology and Aging*, 12(1): 3–11.

Sörensen, S., Martin Pinquart, D., & Duberstein, P. (2002) "How effective are interventions with caregivers? An updated meta-analysis." *The Gerontologist*, 42(3): 356–72.

Stolee, P., Borrie, M.J., Esbaugh, J, Wagg, J., Woodbury, M.G., Petrella, R.J., & Connidis, I.A. (2001) "Maintenance of functional and quality of life outcomes in geriatric rehabilitation." *The Gerontologist*, 47(suppl. 1): 364.

Stolee, P., Rockwood, K., Fox, R.A., & Streiner, D.L. (1992) "The use of Goal Attainment Scaling in a geriatric care setting." *Journal of the American Geriatrics Society*, 40: 574–8.

Stolee, P., Stadnyk, K., Myers, A.M., & Rockwood, K. (1999) "An individualized approach to outcome measurement in geriatric rehabilitation." *Journal of Gerontology: Medical Sciences*, 54A: M641–M647.

Stolee, P., Zaza, C., Pedlar, A., & Myers, A.M. (1999) "Clinical experience with Goal Attainment Scaling in geriatric care". *Journal of Aging and Health*. 11(1): 96–124.

Tomoeda, C.K. (2001) "Comprehensive assessment for dementia: a necessity for differential diagnosis." *Seminars in Speech and Language*, 22(4): 275–88.

Turner, S.A. & Street, H.P. (1999) "Assessing carers' training needs: a pilot inquiry". *Aging and Mental Health*, 3(2): 173–8.

Williamson, G.M & Schulz, R. (1993) "Coping with specific stressors in Alzheimer's disease caregiving." *The Gerontologist*, 33: 747–55.

10
Writers with Dementia: the Interplay among Reading, Writing, and Personhood

Ellen Bouchard Ryan, Hendrika Spykerman, and Ann P. Anas

> *Although my body may still be sputtering along, the day will come when I can no longer write a clear sentence and tell a coherent story. That day will be the actual time of death. The person in me who lives on until natural death occurs is only a shadow left by the deadly laugh of Alzheimer's.*
>
> (DeBaggio 2002: 117)

> *I used to love to read science articles and thick literature novels and non-fiction. I find reading difficult now, but think it is important to keep these skills up. And so I try to read simple stories with clear descriptions that produce strong pictures in my mind, or writing that is very beautiful or humorous.*
>
> (Truscott 2003: 15)

People with dementia have eloquently described their experiences with written language. Most professional discussions of Alzheimer's disease and related dementias mention losses in reading and writing skills, while simultaneously commenting that maintaining these activities is good exercise for the brain. In this chapter, we consider how people with dementia make use of reading and writing activities to enhance memory, satisfaction, self-esteem, and interpersonal communication. Then we review empirical findings and lived experiences concerning declines in reading and writing abilities during the progression of Alzheimer's disease and other related dementias. Strategies for coping with these losses are featured next. We conclude by emphasizing the learning, courage, and creativity involved in maintaining one's sense of self in dementia and the role of reading and writing in that process.

We have taken our lead from writers with dementia by incorporating numerous quotations from their published memoirs. We have also cited published selections from a manuscript prepared by one individual who had been a writer (Bill, in Snyder, 1999).

Benefits of reading and writing activities for the person with dementia

Written communication offers a number of benefits beyond oral communication to the person with dementia. Reading is used to gather information about the disease and about how others cope. Information-seeking can provide a sense of control over the threats posed by the disease.

As noted by Truscott earlier, reading is also a major leisure activity for learning, remembering, and enjoyment. Reading for leisure is especially valuable as other roles diminish.

> I can still read silently very quickly – not quite so super abnormally fast as I used to, but still a lot faster than many other 'normal' people. So the world of books is still very accessible to me, a wealth of enjoyment to dip into and savour." (Boden 1998, p. 76).

Writing is a major compensation strategy for persons with dementia. Writing offers possibilities for remembering, sharing and safeguarding memories, and for making a contribution to one's family and society (Dienstag 2003; Weinstein and Sachs 2000). One can compensate for loss of memory and recognition with lists, notebooks, diaries, calendars, and labels for items whose name and function are not always accessible. "Without a shopping list, it is pointless me venturing to the shop. Without my diary, I don't remember what day it is, what anyone is doing, where they are and so on. I don't seem to have space in my brain for that sense of 'Thursday-ness'.... or 'April-ness' or '1998-ness' " (Boden 1998: 62).

Interpersonal communication with written language, including notes, letters, and email, allows for time (and help) to construct each response:

> I have started to communicate with the outside world by e-mail.... Like many other people with early stage dementia, I have a network of friends in that worldwide community, and I keep in touch with a regular exchange of chatter, humor, coping hints, and encouragement.... The advantage of using text, as many of us agree, is that we can write as ideas come and keep coming back to re-edit until we are satisfied about the context and content of our messages. (Truscott 2003: 15).

Writing in a journal and/or writing poetry provides an avenue for self-expression that can then assist oral expression. Like many other older people, individuals with dementia can gain much satisfaction from reminiscing, recording, and sharing their memoirs. As well, writing poetry can facilitate the expression of feelings.

Finally, a few especially articulate individuals have written about their dementia experiences in published works, newsletters, and increasingly on personal websites to assist others with this condition and their loved ones.

Dementia-related losses in reading and writing

Age-related vision problems in general are likely to lead to the need for larger print and specific color contrasts. Compared to normal older adults, the person with dementia is also subject to irregular eye movements, slowed visual processing, and visuospatial perception deficits (Bayles & Kaszniak 1987; Binetti *et al.* 1998; Lueck *et al.* 2000). Persons with mild to moderate dementia have difficulty discriminating words written in different fonts and photographs of objects in different orientations (Glosser *et al.* 2002). Due to these obstacles, individuals with dementia can benefit from text in print that is large, clear, and in a simple font (e.g. Arial) with increased spacing between lines, paragraphs and ideas.

> I read along a line but then when I go to the next line, I can't find it. It takes so long to find the next line that I have forgotten what was on the previous line. (Bill, in Snyder 1999: 49)

Word recognition declines slowly in most forms of dementia and shows a pattern of increasing difficulty with "exception words," words which do not follow regular letter–sound correspondences (e.g., *height, aisle, sew*). Response times for reading both the high and low frequency regular and exception words increase in early stage dementia, but actual word reading errors do not occur until the moderate stage when semantic memory has significantly deteriorated (Bayles & Kaszniak 1987; Strain *et al.* 1998).

Reading comprehension is relatively well-preserved until the mid-stage of dementia when it can be seriously affected (Bayles & Kaszniak 1987; Schmand *et al.* 1998). However, in the early stages, greater difficulty is shown under specific circumstances such as inhibiting information that has been partially activated (Duchek *et al.* 1998). While reading aloud can be useful to focus attention, oral reading requires the added physical movement of speaking.

> I can still read silently very quickly.... But I can't get my mouth to work at normal speeds, when reading out aloud. (Boden 1998: 77).

Writing deficits have been linked to stages of dementia for many aspects of performance, from the surface level of handwriting to the more complex levels of narrative and informational content (Bayles & Kaszniak 1987; Hughes *et al.* 1997; Kemper & Mitzner 2001). Handwriting declines progressively in dementia (Slavin *et al.* 1999).

> I am having trouble reading the writing I do with a pencil or pen. It used to be sharp; now it wobbles and is full of uncertainty. The words come normally but the letters are sometimes not in proper order. (DeBaggio 2002: 35).

Typing is easier, but it, too, eventually deteriorates.

> I can hardly type anymore. I used to do all of my typing throughout my career... (Bill, in Snyder 1999: 42).

The spelling of words becomes gradually more dependent on knowledge of letter–sound correspondences as lexical memory becomes less accessible (Croisile *et al.* 1996; Luzzatti *et al.* 2003):

> even then, there are those awful moments when I struggle to remember whether it is "right" or "rite", "there" or "their" – very scary for someone who was always so critical of others' spelling and grammar (Boden 1998: 76).

The quality of writing declines, especially in terms of pragmatics, idea density, grammatical complexity, diversity and accuracy of vocabulary, punctuation, organizational structure, and spontaneity (Bayles & Kaszniak 1987; Hughes *et al.* 1997; Kemper & Mitzner 2001; Kemper *et al.* 2001).

> Words come when I sit down to write, but they dance away seductively... (DeBaggio 2002: 31).
>
> With failing memory, it is difficult to write long passages without getting lost in words. Where does the story go? (DeBaggio 2002: 57).
>
> Writing is laborious and time-consuming! (Truscott 2003: 15).

The dementia writers have documented how they learned to cope with written language difficulties. Their strategies, shown in Table 10.1, vary from preservation of energy, repetition, simplification, and taking notes, to the judicious help from others and use of technology. Despite their specific reading and writing deficits and their memory problems, they

Table 10.1 Strategies for dealing with reading and writing losses

Preserving energy	"I have to be careful not to overuse the 'brain battery', and so I have to . . . pick activities of high interest and high value to me." (Truscott 2003: 15).
Repetition	". . . and then to keep re-reading parts until I have a firm hold of the plot, and to read in as large chunks as possible so that I haven't forgotten too much of the plot by the time I return to reading once more." (Boden, 1998: 82). "I make mistakes all the time. I was an editor and I can't remember what I want to write or how I'm going to do it. I have to go back and rewrite again and again." (Bill quoted in Snyder 1999: 42).
Choosing simpler/different materials	"Newspaper articles are easier – short and simple, and much of the information is not important to recall. I check the headlines, picking out items that will interest me." (Truscott 2003: 15). "Because I can no longer read, ... I am eligible for talking books for the blind, and these are a great comfort to me when my mind needs something to occupy it" (Davis 1989: 93).
Simplifying environment	"With books, I need an absolutely quiet environment (not always easy with daughters in and out of the house) . . ." (Boden 1998: 81).
Taking notes as memory aids	"After missing one or two appointments, I began to carry a calendar and note my appointments. I even began carrying slips of paper to remind myself of certain things." (Davis 1989: 95). "If the story is complex, I will make out a list of people in the story and their roles, to minimize flipping back and forth through the pages to remember what's happening." (Truscott 2003: 15).
Help from others	"To write a speech or an article such as this, I have to go through numerous short spurts of writings and re-edit many times, and then ask for outside opinions on the text to complete final edits." (Truscott 2003: 15). "This act of relinquishing to her has relieved me of many of the things that would rob me of peace." (Davis 1989: 105).

Technology	". . . . thoughts appeared much more completely in the typewritten material. In addition to having less mistakes, it was easier for me to see any omissions." (Davis 1989: 97). "I have suggested to many people that they try a handheld recording device, speak into it whenever ideas or messages strike them, and then have someone else type it out for them." (Truscott 2003: 15). "Nothing must happen to 'my' friend, the word processor. It acts as my memory. It spells for me. It even corrects my grammar . . . sometimes. It never tells me to speak faster or slow down. It permits me time to think. It permits me corrections." (McGowin 1993: 124–5).

have persisted in finding often creative solutions around these obstacles. The quotations at the beginning of this chapter highlight the passion with which persons with dementia can be driven to maintain highly valued activities (DeBaggio 2002) and the importance of setting individualized priorities (Truscott 2003).

Personhood, dementia, and written communication

By carefully reading the words of people with dementia about the benefits and challenges of written communication, we have been moved to a more optimistic view of the possibilities for continuing to affirm personhood through reading and writing activities (Harris 2002; Kitwood 1997; Kitwood & Bredin 1992). Reading and writing afford opportunities to cope with one's illness, maintain self-esteem, connect with others, continue learning, enjoy stories, express oneself creatively, and support others.

> As difficult as it is for me to read and write any more or talk about myself, I think it is important therapy for me and a help to others to know what I am experiencing (Bill, in Snyder 1999: 35).

Most certainly, these published authors are exemplary as persons, writers, and sufferers with dementia. However, their writings illuminate objectives and pathways for others who may need more guidance. They provide hope concerning the extent of new learning, creativity, personal growth, and detailed insight available even in the midst of otherwise overwhelming losses.

Once reading and writing activities are established as desirable and possible, a variety of approaches have been offered to facilitate them. Especially for individuals who previously were avid readers or active letter writers, providing just a few missing steps or incentives to get past the inertia of dementia can lead to rewarding successes in valued activities (Dawson *et al.* 1993; Ronch & Goldfield 2003; Souren & Franssen 1993). A quiet environment with good lighting is a vital initial condition (Santo-Pietro & Ostuni 2003). For easy writing and reading of what one has written, marker pens, large lined paper, journals, and large print calendars can be especially helpful. Reading can be assisted as needed with large font size, simple font, good amount of white space, familiar material, familiar words, and group discussion (Bourgeois 2001; Stevens *et al.* 1993). Talking books can be a valuable alternative for reading extended materials, but note that there may come a time when taped material is only understood when in a familiar voice (Davis 1989). Writing can be facilitated with use of the computer, group reminiscence, facilitated group poetry, remembering boxes, and writing partners (Hagens *et al.* 2003; Killick 2003).

We end with a poem by one of our guides.

MISSING TUNIS – STORM IN THE MEDITERRANEAN

Glowing, growing
Flowing, blowing
Growling, scowling
The waves reach higher
Foaming and frothing
Shouting their anger
Tossing, twisting
Shifting, drifting
Our ship sways
As the storm carries us along

Lifted high
Slammed down
Water rushes over
The bow drowns
The ship creaks
Groans, moans
Through the night
The long, dark night
Morning the sea is still
But we've changed course
Missed our port

Acknowledgments

The authors express their appreciation for partial support for this research by a grant from the Social Sciences and Humanities Research Council of Canada.

The authors wish to thank Marilyn Truscott, Michelle Bourgeois, and Marie Savundranayagam for their assistance with this chapter.

References

Bayles, K.A. & Kaszniak, A.W. (1987) *Communication and Cognition in Normal Aging and Dementia*. Boston: Little, Brown.

Binetti, G., Cappa, S.F., Magni, E., Padovani, A., Bianchetti, A., & Trabucchi, M. (1998) "Visual and spatial perception in the early phase of Alzheimer's disease." *Neuropsychology*, 12: 29–33.

Boden, C. (1998) *Who Will I Be When I Die?* Australia: HarperCollins.

Bourgeois, M.S. (2001) "Matching Activity Modifications to the Progression of Functional Changes", in E. Eisner (ed.), *"Can Do" Communication and Activity for Adults with Alzheimer's Disease: Strength-based Assessment and Activities* (pp. 101–7). Austin, Texas: Pro-Ed.

Croisile, B., Brabant, M.J., Carmol, T., Lepage, Y., Aimard, G., & Trillet, M. (1996) "Comparison between oral and written spelling in Alzheimer's disease." *Brain and Language*, 54: 361–87.

Davis, R. (1989) *My Journey into Alzheimer's Disease*. Wheaton, IL: Tyndale House.

Dawson, P., Wells, D.L., & Kline, K. (1993) *Enhancing the Abilities of Persons with Alzheimer's and Related Dementias: A Nursing Perspective*. NY: Springer.

DeBaggio, T. (2002) *Losing My Mind: An Intimate Look at Life with Alzheimer's*. NY: Free Press.

Dienstag, A. (2003) "Lessons from the Lifelines Writing Group for People in the Early Stages of Alzheimer's Disease: Forgetting that We Don't Remember", in J.L. Ronch & J.A. Goldfield (eds.), *Mental Wellness in Aging: Strengths-based Approaches* (pp. 343–52). Baltimore: Health Professions Press.

Duchek, J.M., Balota, D.A., & Thessing, V.C. (1998) "Inhibition of visual and conceptual information during reading in healthy aging and Alzheimer's disease". *Aging, Neuropsychology, and Cognition*, 5(3): 169–81.

Glosser, G., Baker, K.M., de Vries, J.J., Alavi, A., Grossman, M., & Clark, C.M. (2002) "Disturbed visual processing contributes to impaired reading in Alzheimer's disease". *Neuropsychologia*, 40: 902–9.

Hagens, C., Beaman, A., & Ryan, E.B. (2003) "Reminiscing, poetry writing, and remembering boxes: personhood-centered communication with cognitively impaired older adults". *Activities, Adaptation, and Aging*, 27 (3/4): 97-112.

Harris, P.B. (ed.) (2002) *The Person with Alzheimer's Disease: Pathways to Understanding the Experience*. Baltimore: Johns Hopkins University Press.

Hughes, J.C., Graham, N., Patterson, K., & Hodges, J.R. (1997) "Dysgraphia in mild dementia of Alzheimer's type". *Neuropsychologia*, 35: 533–45.

Kemper, S. & Mitzner, T.L. (2001) "Language Production and Comprehension." In J.E. Birren & K.W. Schaie (eds.) *Handbook of the Psychology of Aging* (5th Edition) (pp. 378–93). San Diego: Academic Press.

Kemper, S., Thompson, M., & Marquis, J. (2001) "Longitudinal change in language production: effects of aging and dementia on grammatical complexity and propositional content". *Psychology and Aging*, 16: 600–14.

Killick, J. (2003) "Memorializing dementia". *Alzheimer's Care Quarterly*, 4(1): 18–25.

Kitwood, T. (1997) *Dementia Reconsidered: The Person Comes First*. Philadelphia: Open University Press.

Kitwood T. & Bredin K. (1992) "Towards a theory of dementia care: personhood and well-being". *Aging and Society*, 12: 269–87.

Lueck, K.L., Mendez, M.F., & Perryman, K.M. (2000) "Eye movement abnormalities during reading in patients with Alzheimer's disease". *Neuropsychiatry, Neuropsychology, and Behavioral Neurology*, 13(2): 77–82.

Luzzatti, C., Laiacona, M., & Agazzi, D. (2003) "Multiple patterns of writing disorders in dementia of the Alzheimer type and their evolution". *Neuropsychologia*, 41: 759–72.

McGowin, D.F. (1993) *Living in the Labyrinth: A Personal Journey through the Maze of Alzheimer's*. NY: Delacorte Press.

Ronch, J.L. & Goldfield, J. (eds.) (2003) *Mental Wellness in Aging: Strengths-based Approaches*. Baltimore: Health Professions Press.

Santo-Pietro, M.J. & Ostuni, E. (2003) *Successful Communication with Persons with Alzheimer's Disease, An In-service Manual*. (2nd Edition) St. Louis, MO: Butterworth-Heinemann.

Schmand, B. Geerlings, M.I., Jonker, C., & Lindeboom, J. (1998) "Reading ability as an estimator of premorbid intelligence: does it remain stable in emergent dementia?". *Journal of Clinical and Experimental Neuropsychology*, 20(1): 42–51.

Slavin, M.J., Phillips, J.G., Bradshaw, J.L., Hall, K.A., & Presnell, I. (1999) "Consistency of handwriting movements in dementia of the Alzheimer's type: a comparison with Huntington's and Parkinson's diseases. *Journal of the International Neuropsychological Society*, 5(1): 20–5.

Snyder, L. (1999) *Speaking Our Minds: Personal Reflections from Individuals with Alzheimer's*. San Francisco: W.H. Freeman.

Souren, L. & Franssen, E. (1993) *Broken Connections: Alzheimer's Disease, Part II – Practical Guidelines for Caring for the Alzheimer Patient*. Berwyn, PA: Swets and Zeitlinger.

Stevens, A.B., King, C.A., & Camp, C.J. (1993) "Improving prose memory and social interaction using question asking reading with adult day care clients". *Educational Gerontology*, 19: 651–62.

Strain, E., Patterson, K., Graham, N., & Hodges, J.R. (1998) "Word reading in Alzheimer's disease: cross-sectional and longitudinal analyses of response time and accuracy data". *Neuropsychologia*, 36: 155–71.

Truscott, M. (2003) "Life in the slow lane". *Alzheimer's Care Quarterly*, 4(1): 11–17.

Weinstein, C.S. & Sachs, W. (2000) "Memory 101: a psychotherapist's guide to understanding and teaching memory strategies to patients and significant others". *Journal of Geriatric Psychiatry*, 33: 5–26.

11

Simulating Alzheimer's Discourse for Caregiver Training in Artificial Intelligence-based Dialogue Systems

Nancy Green

Overview

A relatively new focus of computer science research in artificial intelligence (AI) is the development of *embodied conversational agents* (ECAs) (Cassell *et al.* 2000a; Johnson *et al.* 2000). Created by a computer program, an ECA is an animated character whose modes of communication may include speech, gesture, and facial expressions. The words uttered by an ECA may be scripted by a human author, or partially or completely synthesized from more abstract representations (e.g. of speaker goals and affect and propositional content) via natural language generation techniques (Reiter & Dale 2000). Similarly, intonation, gesture and facial expression may be scripted, or generated by the computer program based upon functional models of paralinguistic communication (Cassell *et al.* 2000b). ECAs can be programmed to react to sensed or simulated events in their environment, or to respond to spoken or typewritten language provided by a computer user. AI models of emotion and personality may be used to enhance an ECA's believability (Elliott & Brzezinski 1998). The motivation for developing ECAs is two-fold. First, an ECA can be used to submit a linguistic or psychological model to testing, i.e., via simulation instead of testing with human subjects. Second, ECAs provide a new and compelling mode of human–computer interaction for education and entertainment. Compared to conventional modes of human–computer interaction, benefits may include higher user engagement and motivation and better transfer of skills to real-life situations (Lombard & Ditton 1997).

The goal of our research is to develop ECAs for simulating social conversation between persons affected by Alzheimer's Disease (AD) and their caregivers (Green 2002; Green & Davis 2003; Green 2004). In the long-term, the motivation is to construct testable computational models that will enhance understanding of the effects of AD on communication. However, our immediate goal is to develop a computer program for training caregivers (family members and healthcare workers) in use of assistive techniques for social conversation with persons in their care affected by AD. This training system will enable a caregiver to converse, through an ECA depicting a caregiver (called the caregiver's *Avatar*), with another ECA simulating a person affected by AD (called the *AD Agent*, or *ADA* for short). At certain points in the simulated conversation, the caregiver will be given a chance to select the Avatar's next dialogue action from a list of choices, then will see and hear the Avatar perform the action, and witness its effect on the ensuing conversation. We hope that by having the opportunity to practice assistive dialogue techniques in a simulated conversation the caregiver will acquire techniques that transfer to conversational interactions with persons in his care. Cognitive-linguistic stimulation can improve or maintain the functioning of Alzheimer's dementia patients (Mahendra 2001). Learning how to converse more effectively with persons in their care may reduce caregiver stress as well.

One training objective is to help the caregiver to recognize characteristics of AD discourse, including typical coping strategies employed by AD speakers. For example, AD speakers often use colloquial or figurative language to solicit topic closure (Davis *et al.* 2000). A related goal is to teach the caregiver assistive techniques for responding to such difficulties, e.g., by initiating an affirm–confirm–extend topic closure routine (Drew & Holt 1998) after the AD speaker's bid to close a topic (Davis *et al.* 2000). Another objective is to teach the caregiver techniques for keeping a conversation going by enabling the AD speaker to contribute whatever she can. For example, by use of a request for confirmation (e.g. *Your daughter gave you that, didn't she?*) in place of a request for information (e.g. *Who gave you that?*), the caregiver can make it easier for the AD speaker to provide an appropriate response (Moore & Davis 2002).

Another objective is to enable the caregiver to help an AD speaker to reminisce. Gerontological research has discussed the psychological benefits to the elderly of recounting their autobiographical stories (Mills & Coleman 1994; Golander & Ras 1996). Training to support this activity includes learning to recognize story elements and learning story elicitation techniques. While, to some, the ability to tell an autobiographical story may seem to be beyond the capability of a typical person affected by AD, Shenk *et al.* (2002) have found that co-constructed stories of elderly persons with AD are similar in many respects to stories told by elderly persons without AD. For example in one recorded conversation, Glory (a speaker with AD from a rural background) has difficulty listing all the items that were grown on her childhood farm. On the other hand, from a narratological perspective her story includes narrative structural components (orientation, complicating action, resolution, and evaluation) and uses a variety of linguistic devices to make her story's point. Also, note that just as cultural differences in narrative conventions can lead to "misdiagnosis of cultural difference as deficit" in education (McCabe & Bliss 2003), these differences may impede recognition of story elements by a caregiver. Thus, another goal of the system is to provide training to healthcare workers whose sociocultural backgrounds differ from their clients' backgrounds.

The potential role of AI in AD caregiver training

Currently researchers are studying how AI technologies involving robotics, automated sensing, and automated prompting can be used to help the elderly in daily life (Haigh 2002). Also, several healthcare-related systems using ECAs have been developed. For example, Carmen's Bright Ideas (Marsella *et al.* 2003) is an instructional system to allow mothers of pediatric cancer patients to improve problem-solving skills by participating in a simulated counseling session. Cavalluzzi *et al.* (2003) developed an ECA that provides a user with advice on nutrition. Hubal *et al.* (2003) developed an ECA for training police officers for encounters with the mentally ill.

There are three main potential roles of AI in our proposed AD caregiver training systems. First, AI can be used to partially or completely control the ECAs, as an alternative to following a complete set of instructions provided by a human designer specifying every aspect of their behavior, including what to say and how to say it. The ability to go beyond completely scripted behavior should contribute towards creating believable characters by providing variation and unpredictability (Loyall 1997). Also, it should enable new characters and new "lessons" to be created with much less effort once the AI computer programs have been developed. (This is analogous to the situation whereby it costs less to write a script for human actors than to animate a film frame by frame, since human actors do not need to be told how to perform basic human actions, and moreover they can use their intelligence to improvise and express their interpretation of a character.) Another potential practical benefit of using AI to generate the ECA's language is that it should enable the system to produce dialogues in multiple languages without requiring human translators (Calloway *et al.* 1999).

In addition, use of AI to control the ADA's language can enable models of the effects of AD on communication to be tested through simulation. For example, one part of the model could simulate delays in lexical access from long-term memory; these delays could in turn affect the natural language-generation component, resulting in use of inappropriately vague words in the ADA's speech (Green & Davis 2003). Furthermore, a parameterized generation model could be designed to simulate the progression of AD or individual differences in the effects of AD.

Second, AI can be used to interpret and/or analyze the caregiver's participation in the conversation and thereby inform the training process. For example, some ECA researchers have investigated AI techniques (from the subfields of speech recognition and natural language interpretation) to enable a computer user to interact with an ECA as if speaking to a real person (Swartout *et al.* 2001; Cassell *et al.* 2000a, 2000b). (Because our research focus is on simulating the AD speaker's side of the conversation, we plan to avoid the quite challenging issues in interpreting unrestricted user input by limiting the user to interaction through her Avatar.) Regardless of whether AI is used to *interpret* the user's input, AI also can be used to *analyze* the user's "state of mind". For example, AI has been used to infer a user's affective state based on biosignals and patterns of behavior (Conati *et al.* 2003), as well as from her language itself (Qu *et al.* 2004). Also, in intelligent tutoring systems research (e.g. Johnson *et al.* 2000), analytical techniques from AI are used to track a student's learning by comparing his mistakes to a model based on learning problems characteristic of the subject matter, and tailoring the lesson accordingly.

The third potential role of AI in developing the caregiver training systems is for automated analysis of transcribed recordings of conversations with persons affected by AD. AI techniques (from the subfields of machine learning, data mining, and text mining) can be used to gain knowledge of which interaction techniques are effective in the recorded conversations. For example, AI techniques could be employed to determine the effect of a candidate intervention on measures such as the ensuing affective state of the AD speaker or the number of turns completed by the AD speaker on the current topic following the other speaker's use of the intervention.

AI in initial proposed system

Initially, our proposed system will use AI to control three aspects of the ECAs' behavior: dialogue management, emotion-personality models, and autobiographical story generation. In dialogue systems using AI for natural language generation, it is useful to distinguish dialogue management decisions (i.e., *what to say*, and *when to say it*) from syntactic and lexical realization decisions (i.e., *how to say it*) (Reiter & Dale 2000). Dialogue management includes turn regulation and meeting discourse expectations (such as the expectation that the current speaker should answer the preceding speaker's question) by selection of context-appropriate discourse actions. (In contrast, realization decisions include selection of lexico-syntactic forms for semantic and pragmatic effects.) As outlined in Green & Davis (2003), a computational model integrating normal (non-AD-affected) speakers' pragmatic routines with coping strategies characteristic of AD discourse could be used for dialogue management in the system.

Emotion-personality models of each ECA will be included because of the important role of emotion and personality in social conversation (Bickmore & Cassell 2000). For example, different conversational actions on the part of the caregiver's Avatar may be appropriate when the ADA appears happy and alert as opposed to when he appears depressed or confused. In the personality model, we include sociolinguistic variables, such as age, gender, and culture, that may influence the speaker's conversational style. Giving different personalities to ECAs is another way to tailor the system for different audiences.

Story generation will be performed incrementally, since the ADA's storytelling actions must be integrated into the conversation. As outlined in Green (2002), our computational model will include the ADA's autobiographical memory (called her *Life Episodes Model*), which is similar to the concept in narratology of the *fabula* (Calloway *et al.* 1999), a chronologically and causally ordered sequence of events on which a story is based. In our proposed system, different Life Episodes Models may be provided by system designers for different ECAs; this is yet another way in which audience-tailored ECAs can be created. AI will be used to generate different stories from a given Life Episodes Model, depending on the speaker's goals, e.g. to illustrate her self-identity (Shenk *et al.* 2002).

As a first step towards creating this system, we currently are implementing a research prototype that uses a simple AI model of emotion and a hard-coded "script" for dialogue management and story generation to control the Avatar and ADA (Green 2004). At the same time, we have begun a qualitative analysis of affect in transcribed recordings of conversations with persons affected by AD, the results of which will inform the development of the ADA's emotion model. In Figure 11.1, we illustrate part of a script and the ADA's emotion model.

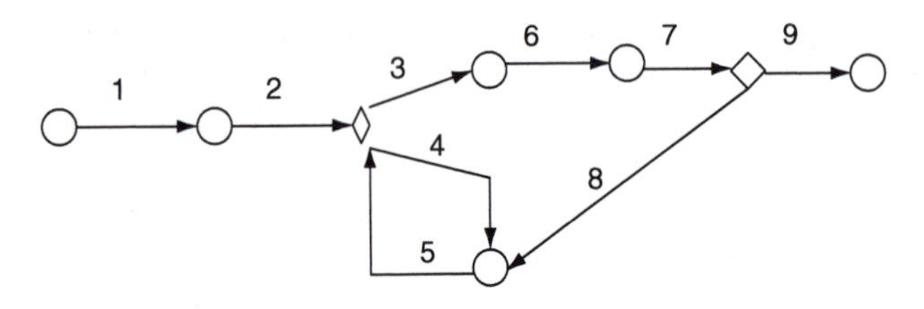

Arcs are described in the table below the graph. Diamond-shaped nodes represent states where the user is given choice. In the table, CA is the caregiver character, ADA is the character with AD.

Arc label	*Agent-precondition*	*Agent-act*	*Agent-effect*	*Beneficiary-effect*
1		CA: Normal greeting ("Hello Mrs. A")		
2	ADA is depressed	ADA: Bare greeting ("Hi")		Add CA belief: ADA seems depressed
3	CA goal: cheer up ADA	CA: Show regard ("It's so good to see you again!")		Increase ADA happiness
4		CA: Direct request story ("Tell me about when you were growing up on the farm.")		Decrease ADA happiness
5	ADA is depressed	ADA: silence		
6	CA goal: cheer up ADA	CA: Story probe ("You were telling me about growing up on a farm . . your family raised corn and cotton . .")		Increase ADA happiness
7	ADA is not depressed	ADA: Open story ("One day I picked over a 100 pounds! And me so small!)		
8		CA: Direct question ("How old were you then?")		Decrease ADA happiness
9		CA: Affirmation ("Alright!")		Increase ADA happiness

Figure 11.1 Part of graph script for caregiver training application (Green 2004)

The graph represents the beginning of a conversation in which, if the caregiver is successful, the ADA will relate an episode from her past. (For readability, the graph's arcs are labeled with row indices for the table below the graph; each row of the table contains conditions, actions, and effects for the corresponding arc.) At the start of this conversation, only one action is possible: the caregiver avatar (CA) gives a greeting. Note that the acts are described in the figure in terms of their dialogue function; examples of utterances that could be used to perform the corresponding function are given in parentheses. Next on arc 2, the response by the ADA is a minimal greeting, presumably because the ADA is depressed. The effect of this act on CA is to add the belief that ADA is depressed. The diamond-shaped node at the end of arc 2 indicates that the caregiver may be given a choice of the actions on arc 3 or arc 4 at this point; the action on arc 3 would be preferable. In contrast, a direct question, as specified on arc 4, would be met with silence. In order to make the preferable choice in this case, the caregiver would need to recognize from the ADA's bare greeting that the ADA may be depressed, and then choose arc 3 as the appropriate response. As shown in the graph, a successful conversation for the purposes of this example would begin with the path (1, 2, 3, 6, 7, 9). If the user chooses arcs 4 or 8, the conversation will suffer.

References

Bickmore, T. & Cassell, J. (2000) "How about this weather? Social dialogue with Embodied Conversational Agents." *Proceedings of the AAAI Fall Symposium on Socially Intelligent Agents.*

Calloway, C., Daniel, B., & Lester, J. (1999) "Multilingual Natural Language Generation for 3D Learning Environments." *Proceedings of the 1999 Argentine Symposium on Artificial Intelligence*, 177–90.

Cassell, J., Sullivan, J., Prevost, S., & Churchill, E. (eds.) (2000a) *Embodied Conversational Agents.* Cambridge, MA: MIT Press.

Cassell, J., Bickmore, T., Campbell, L., Vilhjalmsson, H., & Yan, H. (2000b) Conversation as a system framework: designing Embodied Conversational Agents." In Cassell, J. *et al.* (eds.), *Embodied Conversational Agents.* Cambridge, MA: MIT Press.

Cavalluzzi, A., De Carolis, B., Carofiglio, V., & Grassano, G. (2003) "Emotional Dialogs with an Embodied Agent". In Brusilovsky, P., Corbett, A., & de Rosis F. (eds.) *Proceedings of User Modeling 2003: 9th International Conference.*

Conati, C., Hudlicka, E., & Lisetti, C. (eds.) (2003) "Workshop proceedings: assessing and adapting to user attitudes and affect: why, when and how?," 9th International Conference on User Modeling.

Davis, B., Moore, L., & Peacock, J. (2000) "Frozen phrases as requests for topic management: effect and affect in recepient design by a speaker of Alzheimer's discourse." Presented at *NWAV2000*, Michigan State University.

Drew, P. & Holt, E. (1998) "Figures of speech: figurative expressions and the management of topic transition in conversation." *Language in Society* 27:495–522.

Elliott, C. & Brzezinski, J. (1998) "Autonomous agents as synthetic characters." *AI Magazine*, Summer 1998, 13–30.

Golander, H. & Raz, A.E. (1996) "The mask of dementia: images of 'demented residents' in a nursing ward." *Ageing and Society* 16: 269–85.

Green, N. (2002) "A Virtual World for Coaching Caregivers of Persons with Alzheimer's Disease". *Papers from the AAAI Workshop: Automation as Caregiver: The Role of Intelligent Technology in Elder Care*, 18–23.

Green, N. & Davis, B. (2003) "Dialogue Generation in an Assistive Conversation Skills Training System for Caregivers of Persons with Alzheimer's Disease." *Natural Language Generation in Spoken and Written Dialogue: Papers from the 2003 AAAI Spring Symposium*, 36–43.

Green, N. (2004) "A software architecture for simulated human–human conversation with user intervention." *Proceedings of IEEE SoutheastCon.*

Haigh, K. (ed.) (2002) *Papers from the AAAI Workshop: Automation as Caregiver: The Role of Intelligent Technology in Elder Care.*

Hubal, R.C., Frank, G.A., & Guinn, C.I. (2003) "Lessons learned in modeling schizophrenic and depressed responsive virtual humans for training." In *Proceedings of the Intelligent User Interface Conference.*

Johnson, W.L., Rickel, J.W., & Lester, J.C. (2000) "Animated pedagogical agents: face-to-face interaction in interactive learning environments." *International Journal of Artificial Intelligence in Education.*

Lombard, M. & Ditton, T. (1997) "At the heart of it all: the concept of presence." *Journal of Computer Mediated Communication* 3(2).

Loyall, A.B. (1997) "Believable Agents: Building Interactive Personalities." Unpublished Ph.D. dissertation. Pittsburgh, PA: Carnegie Mellon University.

Mahendra, N. (2001) "Direct interventions for improving the performance of individuals with Alzheimer's disease." *Seminars in Speech and Language* 22(4).

Marsella, S.C., Johnson, L.W., & LaBore, C.M. (2003) "Interactive pedagogical drama for health interventions." In *Proceedings of AIED2003: Conference on Artificial Intelligence in Education.*

McCabe, A. & Bliss, L.S. (2003) *Patterns of Narrative Discourse: a Multicultural, Life Span Approach*. Boston: Pearson Education.

Mills, M.A. & Coleman, P.G. (1994) "Nostalgic memories in dementia: a case study." *International Journal of Aging and Human Development* 38(3): 203–19.

Moore, L. & Davis, B. (2002) "Quilting narrative: using repetition techniques to help elderly communicators." *Geriatric Nursing* 23:2–5.

Qu, Y., Shanahan, J., & Wiebe, J. (eds) (2004) *Working Notes of AAAI 2004 Spring Symposium on Exploring Attitude and Affect in Text: Theories and Applications.*

Reiter, E. & Dale, R. (2000) *Building Natural Language Generation Systems*. Cambridge: Cambridge University Press.

Shenk, D., Davis, B., Peacock, J.R., & Moore, L. (2002) "Narratives and self-identity in later life: case studies of two rural older women". *International Journal of Artifical Intelligence in Education* 11: 47–8.

Swartout, W. *et al.* (2001) "Towards the holodeck: integrating graphics, sound, character and story". *Proceedings of the Fifth International Conference on Autonomous Agents*.

12 Understanding Text about Alzheimer's Dementia

Lisa Russell-Pinson and Linda Moore

Introduction

Managed health care often presents obstacles for both health care providers and patients. Although clinicians are required to manage the increasingly complex health care needs of individuals, managed care usually permits few opportunities or incentives for health care providers to spend the extended time required in explanations of information for patients and family members to completely understand their health care needs. "This development in the way health care is currently delivered has put the onus squarely on the shoulders of patients and their families not only to understand complex treatment choices but also to make informed decisions regarding them" (Russell-Pinson 2002: 5). As a consequence, print and Internet sources are becoming increasingly popular avenues for dispensing health care information for patients and their caregivers (Bauerle Bass 2003; Brody 1999; Fisher 1996; Napoli 1999) and, based on the desires and needs of patients and family members to know more about their own health and disease processes, researchers predict that this trend will continue in the coming years (Bauerle Bass 2003; Kassirer 1995; Parker *et al.* 1999; Sharpe 2000; Zeliff 2001). As Sharpe (2000: 207) points out,

> Today patients can locate information on virtually any illness or condition on virtually any number of Web sites. They can access and search "virtual" medical and nursing libraries, and then compose and propose thorough and meaningful inquiries and comments about their health and care. Well-guided access to information on the WWW can educate patients, reduce their fears, uncertainties, and anxieties, and enhance communication between patients and

> provider. Knowledge and facts can improve attitudes, encourage compliance, provide hope, and improve the prognosis for recovery.

Obtaining health care information from sources other than health care providers can empower patients and enhance outcomes but such educational materials must be approached carefully. Bauerle Bass (2003: 34) cautions practitioners and educators:

> The Internet provides us with new ways to help people become active participants in their health. We must acknowledge, however, that those same people are faced with a quagmire of Internet health information, some correct, some incorrect. Practitioners must assist people in discerning correct from incorrect....

Health care providers can be overwhelmed by sorting through this information, however, because they often do not have the time necessary to assess either the quality of web-based health information content or its presentation during a typical office visit. In addition, even medically accurate materials can pose challenges to health care providers as patients frequently bring resources that they do not fully comprehend to health care providers for further clarification.

Due to the unprecedented accessibility and availability of Internet-based materials produced for patients and the potential problems for both patients and providers that are presented by their use, a new focus on the language of this information has surfaced. It is essential that the language of these materials be investigated and understood for patients and their caregivers "can then be educated on how to better interpret these texts for their own purposes" (Russell-Pinson 2002: 5).

This study focuses on the language of a well-known Internet-based resource on Alzheimer's Dementia (AD). Specifically, it examines the role of interrogatives in this document. The results of this study can be used to expand our understanding of how information about AD is communicated to patients and caregivers, and to develop insights that can improve patient education efforts in both traditional and clinical settings.

Methodology

The Alzheimer's Association is a leading organization for AD research and advocacy. Its website contains extensive information designed to educate those with AD, as well as their caregivers, and is one of the main AD resources cited by health care providers for patients and families in the U.S. Given its prominent role in patient education for AD and its first place ranking for the term *Alzheimer's* in Google (November 21, 2003), a large number of people seeking information on AD on the Internet access this website to obtain the most up-to-date information on AD diagnosis, treatment and news; based on its reputation as a credible and popular resource on AD, the Alzheimer's Association website was chosen as the basis of this analysis.

The Alzheimer's Association website was downloaded on December 15, 2003. A patient education specialist reviewed the website and recommended that its three main sections – "About Alzheimer's," "I Have Alzheimer's," and "Alzheimer's Caregivers" – be examined because these sections contain accurate information and are intended to address a lay readership. The three sections combined contain 21,831 words.

The type of research undertaken in this study is termed *content research,* in that it seeks to assess "the quality of Internet health information" (Bauerle Bass 2003: 23). Such studies are valuable in determining "the types of Internet health information available to consumers . . . to ensure that the best information is available in the best way" (Bauerle Bass 2003: 32–3). Results from content research can be used to establish guidelines for medical websites, including those related to the way that information is presented to patients and caregivers.

The importance of interactivity in patient education by nursing has been recognized since Peplau's (1952) seminal work, *Interpersonal Relations in Nursing*. As Gastmans (1998: 1315–16) notes,

> both the patient and the nurse contribute to and participate in the relational process that unfolds between them. . . . A nurse encourages patients to make their own decisions, so that they can become autonomous individuals, and assists them by providing them the necessary background information, proposing possible alternatives and explaining the consequences of their choices.

A recent development in interactivity has been the presentation of written health information by health care providers to patients and caregivers. Such materials are valuable as an efficient and effective transfer of health information in that they allow readers to control the speed at which they can learn and may be "more difficult to misconstrue than . . . oral language" (Ainsworth-Vaughn 1998: 293; cf. Rankin &

Stallings 2001); additionally, the health information available via the Internet allows access to materials at times and places convenient to users and can enhance the teaching previously done by professionals in clinical settings (Darkin & Cary 2000; Redmond 1997; Sharpe 2000). These printed and web-delivered materials are usually designed by health educators to address the needs of patients and caregivers by presenting health information in ways that take readers' background knowledge into account and to appeal to readers by making texts "*authentically* [emphasis in original] interactional, not a mere description of the symptoms and treatment of the disease," in order to simulate "face-to-face doctor–patient communication" (Al-Sharief 1996: 10). It is this perspective of interactivity in written medical discourse that serves as the basis of this research.

In addition, this study benefits greatly from the literature on *genre analysis* (cf. Bhatia 1993; Berkenkotter & Huckin 1995; Swales 1990). In this tradition, genres – texts that are "highly structured and conventionalised" (Bhatia 1993: 13) – are examined for certain properties, including content, grammar and discourse structure. In genre analyses, *interactivity* refers to textual features, such as first- and second-person pronouns, imperatives and interrogatives, "that exhibit an 'over-signalling' of the reader's responses and reactions" (Al-Sharief 1996: 13) and anticipate reader questions.

The three sections of the Alzheimer's Association website were coded for an interactive linguistic feature: interrogatives. Interrogatives were selected because they "are devices that explicitly address readers, either to focus their attention or include them as discourse participants" (Hyland 2000: 113) and "are the strategy of dialogic involvement *par excellence*, often functioning to express an imbalance of knowledge between participants, but also working to create rapport and intimacy" (Hyland 2002: 530). Furthermore, they have been associated with interactivity in medical writings produced for lay audiences since they "make content accessible [and] encourage the reader to take action" (Russell-Pinson 2002: 265; cf. Al-Sharief 1996).

The coding of the interrogatives was performed by a linguist, in conjunction with a health care provider, thus ensuring reliability. Additionally, collaboration between linguists and subject-area specialists, such as health care providers, "brings validity... and adds psychological reality" (Bhatia 1993: 34) to the findings of genre analyses so both authors analyzed the data independently and conferred on their conclusions.

Results

The examination of the three sections of the Alzheimer's Association website yielded 71 interrogatives. The interrogatives in the website serve multiple purposes (Table 12.1). The sections below detail each of these functions.

Prompts to ask others

One of the main functions of the interrogatives in the Alzheimer's Association website was as suggested questions for the readers to ask others. Specifically, the interrogatives are prompts for those with AD to use with health care providers and potential caregivers and cover a variety of topics, including assessment, diagnosis, treatment, evaluation, and home care. Table 12.2 summarizes the participants in and topics of the questioning in the website.

Suggested questions can be very useful in guiding patients to discuss important issues regarding their healthcare (Gelford 1993). However, the questions directed at the health care providers, such as those in Figure 12.1, are scripted only for those with very mild AD. Those with a more advanced stage of AD could not manage to ask these questions, nor understand the answers to them, for the cognitive impairment that

Table 12.1 Functions of Interrogatives in the Website

Functions
Prompts to ask others
Model language to use with those who have AD
Organizational headers and attention grabbers
Points of reflection

Table 12.2 Participants in and topics of questioning

Proposed questioners	*Addressees*	*Topics*
Person with AD	Health care providers	Tests Symptoms Diagnoses Behavioral changes Medication Clinical trials
Person with AD	In-home caregivers	Qualifications Sponsoring agency Protocol

characterizes AD interferes with language production and comprehension. Rather, the repeated use of a few questions with simple answers assists those with AD in understanding new information, thereby preventing agitation due to over-stimulation (Lonergan 1996). For example, in lieu of *What kind of assessment will you use to determine if the drug is effective?*, the question could be rephrased for the impaired reader as *How will we know if the drug is working*? This question not only is shorter and less complex than the one in Figure 12.1 but also directly involves the questioner through the use of the first-person plural pronoun *we*. The use of *we* indicates that the questioner, the health care providers and perhaps a caregiver will be involved in assessing the effectiveness of an AD medication. Stressing that the persons with AD and their caregivers are important parts of the health equation through the use of inclusive language may help to empower both parties (cf. Sharpe 2000). From a practical standpoint, it is also important to remember that especially in a managed care environment, health care providers have limited time to spend with patients because the demands for increased productivity, measured by number of patients seen, have reduced the time that they can spend on non-acute issues (Darkins & Cary 2000). Thus, if persons with AD and their caregivers believe that all of the questions in Figure 12.1

Questions for your physician

Clear communication between the physician and the patient or caregiver is essential. Ask your physician the following questions when you discuss any treatment options.

- What kind of assessment will you use to determine if the drug is effective?
- How much time will pass before you will be able to assess the drug's effectiveness?
- How will you monitor for possible side effects?
- What effects should we watch for at home?
- When should we call you?
- Is one treatment option more likely than another to interfere with medications for other conditions?
- What are the concerns with stopping one drug treatment and beginning another?
- At what stage of the disease would you consider it appropriate to stop using the drug?

"About Alzheimer's"

Figure 12.1 Questions directed to health care providers

are essential, it is likely that they will need to book several consecutive appointments and recognize that they may need to pay out-of-pocket for these "extra" visits. Furthermore, the number of questions must be related to how much information persons with AD and their caregivers can absorb at any one time.

Model language to use with those who have AD

The data also include interrogatives as specific examples of questions for caregivers to ask or to avoid asking persons with AD. For instance, the website recommends that caregivers say *May I help you*? ("Alzheimer's Caregivers") when a person with AD becomes agitated and proposes that caregivers refrain from directly quizzing the person with AD by posing questions such as, *You know who that is, don't you*? ("Alzheimer's Caregivers"). While interrogatives suggesting that the reader ask others pointed questions about diagnosis, treatment and care have previously been noted in linguistic literature (cf. Al-Sharief 1996), interrogatives suggesting that the reader ask specific questions in an effort to communicate more effectively with people with a medical disorder have not been seen in linguistic examinations of written medical discourse for lay audiences heretofore. This may be because prior linguistic studies have not examined interrogatives in written texts about cognitive impairment. However, in the current research, it is not surprising that language designed to facilitate successful interaction between those with AD and their caregivers is present since comprehension is compromised in AD and noting particular ways to communicate, such as the use of specific language with persons with AD, may assist readers in thwarting potential misunderstandings with those for whom they care.

While language strategies can be useful tools to ease communication between caregivers and persons with AD (Moore & Davis 2002; Small *et al.* 2003), modeling effective question use for AD caregivers is not as straightforward as it seems. For example, Tappen *et al.* (1997) maintain that caregivers' use of both open-ended and close-ended (i.e., yes/no) questions can facilitate interaction with persons with AD, depending on the context of the questioning; likewise, while the work of Small *et al.* (2003) suggests that caregivers' use of yes–no questions, rather than open-ended questions, contributes to effective interaction when speaking to persons with AD, the researchers write that open-ended questions may still have a place in discourse with those with AD. Small *et al.* (2003: 364) remark,

> [T]he use of yes/no questions seems to be very effective in enabling the person with AD to respond successfully. However, even

> though such questions appear to respect the person's autonomy in that they offer a choice, they can also undermine the individual's personhood by limiting their options. Again, one can ask whether it is appropriate to restrict an individual's personhood by limiting their options. We would argue that caregivers should be encouraged to strike a balance between using effective speech accommodations and encouraging the maintenance of existing abilities.

A misunderstanding of how to use questions appropriately may frustrate caregivers' attempts at successful communication, "have negative consequences for the maintenance of existing abilities and/or . . . reduce the independence of the individual" (ibid.).

Organizational headers and attention grabbers

Interrogatives in the Alzheimer's Association website also function as headers. Similar to their use in other medical texts for lay readers (Al-Sharief 1996; Russell-Pinson 2002; cf. Biesenbach-Lucas 1995), these questions preview the information in subsequent paragraphs and allow the author to connect with the audience by posing questions mostly likely on the minds of the readers, as illustrated by the following header taken from the beginning of the website:

What is Alzheimer's?
("About Alzheimer's")

Headers containing both interrogatives and the second-person pronouns *you* and *your* allow the author to pose questions directly to reader. The questions in headers not only highlight the importance of the information but also invite the readers to reflect on their own situations in a more explicit manner, as the three interrogative headers below demonstrate:

Do you visit your physician annually?
Do you accept assistance from others?
Do you talk to others about your feelings?
("Alzheimer's Caregivers")

This use of interrogatives in headers is related to their function as an attention-grabber in the introductory paragraph of a section on maintaining the caregiver's health:

> *Are you so committed to caregiving tasks that you've neglected your own physical, mental, and emotional well-being?*
>
> ("Alzheimer's Caregivers")

It is curious that these headers are the only questions used to orient and capture the interest of readers in the website. Prior linguistic studies of medical writings produced for laypersons indicate that such questions are an important marker of audience since they typically "serve to help the reader understand the organization of the text" (Russell-Pinson 2002: 194; cf. Biesenbach-Lucas 1995) and promote interaction with the audience (Al-Sharief 1996; Russell-Pinson 2002).

Points of reflection

Interrogatives in the website also pose questions upon which readers should reflect. For example, Figure 12.2 presents questions for caregivers to consider before taking future actions, such as before moving a person with AD into the caregiver's home.

In other cases, interrogatives prompt caregivers to reflect on the past. Figure 12.3 illustrates questions designed to get caregivers to assess the causes of behavioral problems in those with AD.

Consider carefully before moving a loved one into your home

The decision to move the person to your home is influenced by many factors. Here are some things to think about before moving the person into your home:

- Does he or she want to move? What about his or her spouse?
- Is your home equipped for this person?
- Will someone be at home to care for the person?
- How does the rest of the family feel about the move?
- How will this move affect your job, family, and finances?
- What respite services are available in your community to assist you?

("Alzheimer's Caregivers")

Figure 12.2 Questions for reflection on AD caregiving

Exploring causes and solutions

It is important to identify the cause of the challenging behavior and consider possible solutions.

Identify and examine the behavior

- What was the undesirable behavior? Is it harmful to the individual or others?
- What happened before the behavior occurred?
- Did something trigger the behavior?

("Alzheimer's Caregivers")

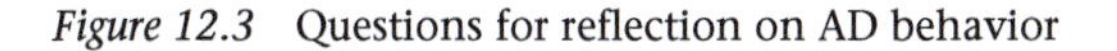

Figure 12.3 Questions for reflection on AD behavior

Questions such as those presented in Figure 12.3 not only serve as points of reflection for the caregiver; in fact, the answers to them can provide valuable information to the health care providers about initial AD staging and symptoms. Furthermore, on-going written responses to questions about different behaviors in an informal journal can present both the caregiver and the health care provider with information about the progression of the disease process as well as the efficacy of the treatment plan.

Discussion

The Alzheimer's Association website provides a valuable resource to those with AD and their caregivers. Because it imparts the most comprehensive, current and accurate information about AD, patients and their families can be confident that their concerns about AD are addressed responsibly and credibly. Also, because it contains separate sections directed at those with AD and their caregivers, the website presents the information most relevant to each audience, which facilitates use.

Since this analysis was conducted, the Alzheimer's Association has updated its website to make it not only more authoritative in terms of medically accepted information, but also more accessible to multiple kinds of users, including health professionals. The revised website still contains many of the above questions and as the association continues to develop its website, it may consider incorporating the following observations regarding interrogatives into its next revision.

Prompts to ask others

Written information on AD should make those with AD and their caregivers aware of how to better communicate with their health care providers, including how to use questions effectively and efficiently during office visits. For instance, the Alzheimer's Association website could include information about how health care providers only have time to address limited health-related issues in any one appointment so that patients and caregivers should try to frame their health concerns in several questions in advance of the visit in order to guide health care providers in presenting the most relevant information about their condition. To illustrate how patients and caregivers can do this, the text could propose three suggested questions, such as *What do you recommend as a treatment?*, *What are the side effects of the drug?* and *How will we know if the drug is working?*, for an appointment mainly focused on starting medications for AD; reducing the number and rephrasing the types of suggested questions may help to ensure that the time during the visit is directed at the immediate issue, that patients and caregivers are not overwhelmed by too much information and that the information is more likely to be retained.

Model language to use with those who have AD

While the revised website still features a question as an example of "quizzing", it no longer contains specific phrases, sentences or questions to model effective language use for caregivers. The inclusion of model language, however, may benefit readers of the "Care Partners" section since certain language strategies can ease communication with those with AD. Because language "strategies should be highlighted in the literature for caregivers" (Small *et al.* 2003: 365) and because the data on the efficacy of different interrogative forms with persons with AD are complex, model questions in patient education texts should be prefaced with brief instructions to focus attention on the questioning strategy in order to explain how to use it appropriately. For example, with respect to question use, Small *et al.* (2003: 364) recommend:

> When it is essential for the person with AD to understand a message, then employing effective simplifying strategies [e.g., yes/no questions] is appropriate. When there is less urgency to the message, caregivers may engage in more typical communication behavior, using appropriate repair strategies as the need arises.

Thus, the website could contain a section on asking questions, one that explains what yes/no and open-ended questions are, provides examples of each type of question and describes when to use each type with a person with AD in order to achieve optimal communication.

Organizational headers and attention grabbers

The Alzheimer's Association website contains strategic use of questions as organizational headers and attention grabbers, which serve to orient and involve readers. However, including more of this feature may further enhance this and other websites. Since interrogatives as headers are central to patient information texts and may be expected by readers, presenting more interrogatives as headers in a specific section, such as "frequently asked questions" (FAQs), in future revisions of the website may be beneficial in not only addressing the most common questions that readers have about AD but also assisting readers to locate this information more easily.

Points of reflection

Currently, the Alzheimer's Association website contains excellent questions upon which caregivers can reflect. The responses to these questions may have more impact to caregivers, however, if the website could encourage readers not only to contemplate the important issues in the questions but also to answer the questions in writing. For instance, before an appointment to assess the efficacy of a medication, persons with AD and their caregivers can review and write answers to a set of questions for reflection about any behavioral changes in the person with AD. Responses to these questions could be used to prompt patients and caregivers about changes in the person with AD when asked by health care providers. This use of questions and answers may help caregivers hone their skills in identifying the triggers of and solutions to challenging behavior, thus empowering them to be more confident in their abilities. The strategy of recording observations may also improve the accuracy of information that health care providers receive about patients' changing conditions so that more immediate intervention can ensue.

Implications

Webber's (1994) comprehensive taxonomy of the function of interrogatives in medical journals indicates that questions in texts produced for health care providers are used differently from those in texts for lay audiences. For example, although both medical journal articles and patient information writings use questions as headers and to pique the interest of the readers, the journal articles also exhibit use of interrogatives as ways to hedge, to criticize, to point to areas for future research and to conclude (Webber 1994: 263–6; cf. Hyland 2002); however, none of the other functions of questions described in the previous section, such as points of reflection and suggested language, has been noted in texts prepared for health care providers.

Since most health care providers use the Internet for up-to-date information on health care management and research rather than health education, they may not be aware of how medical information is communicated to patients online. However, it is imperative that health care providers understand the language used to communicate to patients since "clinicians are being called upon to educate patients on using health-related materials, especially those that are web-based" (Russell-Pinson 2002: 292). As Sharpe (2000: 208) observes,

> During visits, nurses and physicians should ask patients if they use the Internet to search out health care-related information. If the patient is computer literate, the practitioner might consider providing the patient with a list of appropriate Web site addresses. Nurses can prepare lists of such sites and advise the patient and/or their [sic] family to retrieve them or to share and discuss relevant sites they might discover on their own. The ultimate responsibility for educating patients about reliable Internet resources lies with nurses and physicians. They should make every effort to integrate Internet resources into their practice....

For health care providers who treat those with AD, understanding the language of AD, including web-based information about and recommendations for AD care, is crucial because the use of linguistic features, such as interrogatives, in lay texts can introduce, reinforce and enhance successful strategies for communicating with persons with AD and about AD, thus improving health outcomes.

The production of written health materials has traditionally focused on communicating comprehensible content for lay readers by assessing the readability of these materials through measures of word and sentence length and vocabulary (Doak *etal.* 1996; Rankin & Stallings 2001; Redmond 1997). However, as this study indicates, educators preparing materials for patients and caregivers may benefit from also assessing the presence and use of discoursal features such as interrogatives, in texts to further enhance their efficacy. Partnerships among health educators, health care providers and linguists can offer insights that professionals working alone often cannot provide and such collaborations promise the illumination of best practices for developing materials that promote health information retention and ultimately, patient compliance and satisfaction.

Future directions

Few genre analyses have been performed on written patient information; rather, genre analyses have usually been reserved for texts produced for professionals since the main goal of such studies has been to assist non-native speakers to develop awareness of the properties of texts in their disciplines so that they can improve their reading and writing in English (see Bhatia 1993; Dudley-Evans & St. John 1998; Hyland 2000; Swales 1990). This application of genre analyses has been instrumental in the English for Specific Purposes movement, which instructs pre- and in-service professionals in discipline-specific discourse. An offshoot of this movement is English for Medical Purposes (EMP). EMP courses are taken by prospective and practicing medical professionals, such as specialists, general practitioners and nurses, to enhance their English language skills, and genre analyses of texts common to these professions – e.g., research articles, review articles, case reports – are used to inform EMP instruction. For example, Hussin (2002: 29–30) uses genre analysis to teach her English as a second language (ESL) students how to construct "nursing care plans" and other written genres central to the profession.

Due to the importance of written texts to patients' comprehension of medical information, compliance in medical treatment and feeling of empowerment in the medical process (Sharpe 2000), patient information texts should assume a prominent role in future genre analyses and findings from such studies be incorporated into EMP courses so that current and future ESL health care providers can benefit from observations garnered from these analyses; additionally, these findings could enhance courses preparing native English-speaking health care providers in evaluating pre-existing written health information, as well as in designing their own health-related materials, more effectively and efficiently. Professional health literacy educators and patient-materials writers could also gain deeper understandings about the nature of these texts through the results of genre analyses.

Given the dearth of studies on written patient information, we call on multidisciplinary teams to conduct large-scale analyses on medical texts produced for lay readers. In particular, the creation of a corpus – "a large and principled collection of natural texts" (Biber *et al.* 1998: 4) – of this genre could provide further insights on the linguistic features characteristic to these writings. Researchers from varied disciplines, such as linguistics, nursing, gerontology and psychology, will be able to add different perspectives on the findings from such analyses and enrich notions about what constitutes an effective patient information text. Future studies should also include outcomes research, which seeks to uncover how the use of health information affects "the ability to relate to health care providers and . . . perceived self-efficacy to cope with illness" (Bauerle Bass 2003: 23). Unless more comprehensive examinations of patient education materials are conducted and a better understanding of how these texts are interpreted by readers is reached, opportunities for successful patient education may be lost and promising outcomes for patient compliance and satisfaction may be jeopardized.

References

Ainsworth-Vaughn, N. (1998) *Claiming Power in Doctor–Patient Talk*. NY: Oxford University Press.

Al-Sharief, S.M. (1996) "Interaction in Written Discourse: The Choices of Mood, Reference, and Modality in Medical Leaflets." Unpublished MA dissertation. Liverpool, United Kingdom: University of Liverpool. Available on 31 March 1998 at http://www.liv.ac.uk/~ssharief/chapter1.html#top.

Bauerle Bass, S. (2003) How will Internet use affect the patient? A review of computer network and closed Internet-based system studies and the implications in understanding how the use of the Internet affects patient populations. *Journal of Health Psychology*, 8(1): 23–36.

Berkenkotter, C. & T. Huckin (1995) *Genre Knowledge in Disciplinary Communication: Cognition/Culture/Power*. Hillsdale, New Jersey: Lawrence Erlbaum.

Bhatia, V. (1993) *Analysing Genre: Language Use in Professional Settings*. Harlow, UK: Addison Wesley Longman.

Biber, D., S. Conrad & R. Reppen. (1998) *Corpus Linguistics: Investigating Language Structure and Use*. Cambridge: Cambridge University Press.

Biesenbach-Lucas, S. (1995) "A Comparative Genre Analysis: The Research Article and Its Popularization." Ph.D. dissertation. Washington, DC: Georgetown University.

Brody, J.E. (1999, August 31) "Of fact, fiction and medical web sites." *The New York Times on the Web – National Science/Health*. Available on 14 September 1999 at http://www.nytimes.com/library/national/science/083199hth-brody.html.

Darkins, A.W. & M.A. Cary. (2000) *Telemedicine and Telehealth: Principles, Policies, Performance, and Pitfalls*. NY: Springer.

Doak, C., L. Doak & K. Lorig. (1996) "Selecting, Preparing and Using Materials." In K. Lorig *et al.* (eds.), *Patient Education: A Practical Approach* (Second Edition). Thousand Oaks, California: Sage.

Dudley-Evans, T. & M. St. John. (1998) *Developments in English for Specific Purposes: A Multi-Disciplinary Approach*. Cambridge: Cambridge University Press.

Fisher, L. (1996, June 24) "Health on-line: a participatory brand of medicine." *The New York Times on the Web – Technology/Cybertimes*. Available on 14 November 1998 at http://search.nytimes.com/search/daily/b...site+10580+1+wAAA+%22 on-line%7 Ehealth%22.htm.

Gastmans, C. (1998) "Interpersonal relations to nursing: A philosophical-ethical analysis of the work of Hildegard E. Paplau." *Journal of Advanced Nursing*, 29(6): 1312–19.

Gelford, D.E. (1993) *The Aging Network: Programs and Services*. NY: Springer.

Hussin, V. (2002) "An ESP Program for Students of Nursing." In T. Orr (ed.), *English for Specific Purposes*. Alexandria, Virginia: TESOL.

Hyland, K. (2002) "What do they mean? Questions in academic writing." *Text*, 22(4): 529–57.

Hyland, K. (2000) *Disciplinary Discourses: Social Interactions in Academic Writing*. Harlow, UK: Pearson.

Kassirer, J.P. (1995) "Transforming the delivery of health care." *Consumers' Research Magazine* 78: 27–9. Available on 22 October 1997 at http://bart. prod.oclc.org: 3050/FETCH:rec...xt=html/fs_fulltext.htm%22/fstxt114.htm.

Lonergan, E.T. (1996) *Geriatrics: A Lange Clinical Manual*. Stamford, Connecticut: Appleton & Lange.

Moore, L. & B. Davis. (2002) "Quilting narrative using repetition techniques to help elderly communicators." *Geriatric Nursing*, 23(5): 262–6.

Napoli, L. (1999, May 11) "Hospitals reaching new patients online." *The New York Times on the Web – National Science/Health*. Available on 2 August 1999 at http://www.nytimes.com/library/national/science.

Parker, R.M., T.C. Davis & M.V. Williams. (1999) "Patients with Limited Health Literacy." In W.B. Bateman, E.J. Kramer and K.S. Glassman (eds.), *Patient and Family Education in Managed Care and Beyond: Seizing the Teachable Moment*. NY: Springer.

Peplau, H.E. (1952) *Interpersonal Relations in Nursing*. NY: Putnam.

Rankin, S.H. & K.D. Stallings. (2001) "Educational Interventions for Patients and Families." In S.H. Rankin and K.D. Stallings (eds.), *Patient Education: Principles and Practice* (Fourth Edition). Philadelphia, Pennsylvania: Lippincott.

Redmond, B. (1997) *The Process of Patient Education*. St. Louis: C.V. Mosby.

Russell-Pinson, L. (2002) "Linguistic and Extratextual Variation in Medical English Writings: A Comparative Genre Analysis." Unpublished Ph.D. dissertation. Washington, DC: Georgetown University.

Sharpe, C. (2000) *Telenursing: Nursing Practice in Cyberspace*. Westport, Connecticut: Auburn House.

Small, J.A., G. Gutman, S. Makela & B. Hillhouse. (2003) "Strategies used by caregivers of persons with Alzheimer's disease during activities of daily living." *Journal of Speech, Language and Hearing Research*, 46: 353–67.

Swales, J.M. (1990) *Genre Analysis: English in Academic and Research Settings*. Cambridge: Cambridge University Press.

Tappen, R.M., C. Williams-Burgess, J. Edelstein, T. Touhy & S. Fishman. (1997) "Communicating with individuals with Alzheimer's disease: an examination of recommended strategies." *Archives of Psychiatric Nursing*, 11(5): 249–56.

Webber, P. (1994) "The function of questions in different medical journal genres." *English for Specific Purposes*, 13(3): 257–68.

Zeliff, K.S. (2001) "Patient Education Resources on the Internet." In S.H. Rankin & K.D. Stallings (eds.), *Patient Education: Principles and Practice* (Fourth Edition). Philadelphia, Pennsylvania: Lippincott.

13

Epilogue: The Prism, the Soliloquy, the Couch, and the Dance – The Evolving Study of Language and Alzheimer's Disease

Heidi E. Hamilton

Where we've been

I'm writing this epilogue to *Alzheimer Talk, Text and Context* twenty-three years after I began what would become a 4½ year conversational journey with "Elsie," an eighty-one-year-old woman diagnosed with Alzheimer's disease. At the beginning of my conversations with Elsie, I was a twenty-six-year-old graduate student in sociolinguistics, and no stranger to working with elderly individuals. During my years in high school and college I had frequently visited area nursing homes and enjoyed conversing with residents despite the sometimes severe sense-making difficulties that accompanied these talks. My volunteer work at the health care center where Elsie lived involved co-leading weekly armchair exercise and baking classes, assisting with afternoon parties, reading to blind residents, and speaking with a woman who had lost her English abilities and could only speak her first language, German. In my role as participant-observer, I was hoping to supplement the clinical approach taken in most studies of Alzheimer's disease at that time with what Kitwood (1988) has called a "personal research approach":

> The key to a personal approach is that it does not "stand outside," taking the position of a detached and unaffected observer. At its core, it works interpretively and empathetically, going far beyond the measurement of indices or the codification of behaviour. In all of this the researcher takes a personal risk.... It is on the ground of our own experience that we can gain some inkling of what is happening to another. (Kitwood 1988:176)

When Elsie and I began talking, I did not know what I would find – besides friendship. In 1981 when I first got to know Elsie, very little was reported in the scholarly literature about language and Alzheimer's disease. I relied initially on Obler's (1983) insightful review of Irigaray's (1973) study of thirty-two Alzheimer's patients in France as well as on a volume on language, communication, and the elderly edited by Obler & Albert (1980). But as a student of discourse analysis and what would later be called interactional sociolinguistics, it was difficult to find theoretical frameworks and methodologies in the literature that would allow me to capture what I sensed was potentially most significant about Elsie's communication and how it was interrelated with my own.

In the face of the publication of increasing numbers of studies on language and Alzheimer's disease by speech and language pathologists, psycholinguists or formal linguists using speech samples elicited in clinical settings (e.g., Bayles 1979, 1982, 1984; Kempler 1984; Obler 1983; Ripich & Terrell 1988; Ulatowska 1985), I was continually asked – and even came to ask myself! – what a sociolinguistic approach to this problem would look like and indeed, sometimes, whether it was even possible. My resulting study, then, partly by design and partly by necessity, was highly data-driven. Sense-making difficulties and unusual moments in my taped and transcribed conversations would pique my linguistic curiosity along the way and lead me to wonder about possible interrelationships in the data. I looked to solid work in neighboring disciplines and found contextualizing insights in work such as Lubinski (1976, 1981) in communication studies, Boden & Bielby (1986) and Gubrium (1975) in sociology, Coupland *et al.* (1988) in social psychology, and Sabat & Harré (1992) in psychology. But it was not until after the final conversation in my study was recorded that I began to use the analytical tools in my sociolinguistic and discourse analytic tool bag to carry out the qualitative and quantitative analyses which seemed to be true to the data.

To my knowledge, the resulting study reported in Hamilton (1991, 1994a and 1994b) was the first investigation of Alzheimer's disease to examine the language of open-ended, naturally-occurring conversations over time in an attempt to understand not only how communicative abilities and disabilities were related to each other and how they changed over time, but, importantly, how these were influenced by both preemptive and reactive communicative behaviors on the part of the conversational partner. In later work I incorporated notions of face (Goffman 1967), linguistic politeness (Brown & Levinson 1978, 1987), and positioning theory (Davies & Harré 1990) in investigations of the social construction of identities in oral (Hamilton 1996) and in written (Hamilton 2000) communication involving individuals with Alzheimer's disease.

Given that background, I was very pleased to hear from Boyd Davis a little over a year ago about the digital collection of impaired discourse (*Charlotte Narrative and Conversation Collection: Impaired Discourse*) that forms the backbone of the present volume. I was honored to be asked to read the contributions to this book – most of which are based on naturally-occurring conversations with individuals who have Alzheimer's disease – and to write this epilogue, suggesting fruitful next steps in the endeavor to understand language and Alzheimer's disease. In what follows I introduce and briefly discuss four notions – the prism, the soliloquy, the couch, and the dance – that represent major approaches to the study of language and Alzheimer's disease represented both in the field as a whole and within this edited volume specifically. On the backdrop of that discussion, I then offer recommendations for future research at the intersection of language and Alzheimer's disease. The recommendations for basic research are organized within the following four categories: (1) research subjects; (2) contexts of language use; (3) theoretical frameworks; and (4) units of analysis.

Before moving on to that discussion, however, I wish first to provide the reader with two vignettes that are meant to breathe life into the characterizations of language as related to Alzheimer's disease – especially for the reader who has little direct personal experience with this disease. These vignettes are excerpted from Hamilton (2003), my attempt to portray issues of language and Alzheimer's disease in a less technical way than in my earlier work. The interaction represented in the first vignette took place in October 1981 immediately following an armchair exercise class on the 4th floor of Elsie's health care center. The interaction represented in the second vignette took place nearly four years later – in July 1985 – in Elsie's room on the 2nd floor of the center.

Vignette 1: October 1981

Armchair exercise class on the 4th floor of the Center

I carefully lifted the needle directly up off the spinning 33. I returned the arm to its resting place and turned off the record player. The dozen or so men and women seated in our circle of armchairs handed me their plastic hoops, beach balls, and yard-long lengths of ribbon as I walked around, pausing briefly in front of each one. Once the exercise equipment was stowed in the cardboard box in the corner, it was time for us to start our walk down the hallway to the elevators.

Elsie looked in my direction, "Are we to go now? Are we to go now?"

"Yes," I said.

She placed her hands on each arm of her chair and pushed hard as she lifted her weight up off the seat. That accomplished, she leaned down to grab her cane and said, "We'll go now."

A few seconds passed before the next questions came. "Which way do we go? Which way do we go, honey?"

"You'll go around to second," I answered. "You go over this way. We want to catch up with Mrs. Watson. She's already halfway to the elevators." Jill had taken off right away with a small group of more agile residents.

Elsie looked at me again: "You coming with me? You going with me?"

"Yes, uhhuh," I replied.

Elsie seemed absolutely giddy at this news. The fact that she understood what was going on at this moment seemed to energize her. She continued, "Yes. Oh yes. We're . . we're up there on the regular place we step on, don't . . isn't it?"

"That's right," I said. "We just want to go straight to the elevators." I was so focused on my responsibility of getting her home that I had just missed an opportunity to bask with her in the satisfaction of the moment.

"Sure," Elsie continued. "Yes. This is . . this is out main building, so we're . . we're on our regularly time, aren't we, dear honey? Yes. Thank you ever so much." She was truly happy.

As we walked past another female resident, Elsie turned to greet her. "Hello, dear. Glad to see you, honey."

Smiling, she then turned back to me. "Aren't they sweet, honey? Aren't they sweet? We're on our regular place, aren't we? Isn't that wonderful, honey?"

It had only been a few minutes since the end of our exercise class. We still had quite a ways to walk before we caught up with her friend, Mrs. Watson, at the elevators. But for this moment and in this place, Elsie was exhilarated. And who wouldn't be? She recognized where she was and she had just seen some friendly faces. She exuded the sense of calm and sheer happiness that comes from feeling anchored in this world.

(Hamilton 2003: 19–22)

Vignette 2: July 1985
Elsie's room on the 2nd floor of the Center

Elsie had become restless. Her bed was no longer a place for slumber, but had turned into something to be broken out of. She was talking nearly non-stop and I was having more trouble than usual following what she was saying. "And he he wants to get..Yes it will..are the names and kings and kill and we we were trying to get the heart uh their heart and cause they want to trash dead yeah. They have..they're have to get together, dear honey. And they have to get. Listen, dear honey."

Up till then I wasn't even sure she was talking to me. "Yes, what?" I asked.

"I want to get to get. I will have to..hold hold hold hold hold this there." She alternated between looking up at my face and down to her right, toward the siderail on her bed. I looked down and noticed that my left hand was still resting on the bottom rung.

"What?" I asked. "You want me to hold my hand down here?"

"Yes," Elsie responded. "And they said a try try to stay and look at.."

She was jabbing with her right hand at my forearm, right above my wrist.

"This is my arm," I explained. It seemed that she didn't know it was attached to me.

"Really..and then and then I want to go home. And then and then I can I'll pack. I haven't had to tea tea chase..take a call to write that there and I want to know."

Okay, so that much was clear. She wanted to go home.... Something about packing and taking a call. But now she was saying she wanted to know. I asked her, "What do you want to know?"

"Why I," she began. Then she paused. "Let's see. They they said see. They're saying. And pe- pe- crashed there there in the ate. Uh. Now has? Has? Can you? Can?"

We talked like that for a while. I would listen to Elsie and try to grab onto the words that I recognized in her stream of talk. I thought to myself that this was much like the carnival game where one has to stand over rushing water with a net on a stick and try to pluck out the plastic fish before they are swept away.

Sometimes I would try to direct her attention to familiar sights: old photographs or the summery scenery out her window. Sometimes our conversation held its own humor.

At one point I noticed that her nose was runny and I had asked her, "Would you like a Kleenex for your nose?"

Elsie responded, "Mhm. Oh yes. I . . I know his name."

No real connection between what I had just said and what Elsie said, unless – It dawned on me that she was doing the same fishing game as I was. She was grabbing onto the only words she could recognize and trying to make sense of what I had said, incorporating my words into her new turn. When I had said "nose," she had heard "know his" and on we went.

I looked around the room and saw an old issue of *National Geographic* magazine. I thought that might brighten her up. I said, "This is a . . . this is a nice magazine. It's a *National Geographic* magazine. Have you seen this?"

I was surprised by her response. "I don't understand it . . stand it . . and hand." I felt sad. I certainly hadn't meant to make her feel bad.

I asked, "Would you like me to help you look at this magazine?" She didn't respond. I went ahead and opened the magazine up to a bright full-page photograph. I held the magazine in front of her and pointed. "See? This is a man underwater with a big fish."

This wasn't working. The magazine wasn't going to hold her attention today. She was looking around, her eyes darting from one thing to the next. First to my handbag, then to my blouse. I decided to try to talk about the weather, then the 4th of July holiday festivities. "Did you ever go anyplace downtown for the fireworks?" I asked her.

"Well, it seems so," Elsie replied. Her hands and eyes were still restless. I pulled a new tissue from the Kleenex box and leaned over the railing towards her face. Elsie blew lightly into the tissue as I held it for her.

"There you go. Here. Let me take that."

"Uhhuh."

"There you go. Can you breathe better now? That's better." And I leaned over and kissed her cheek.

"Yes," Elsie said softly.

"Mhm."

"Yes."

"So, Elsie. I'll see you later. Okay?"

"Sure," she said.

Then she smiled through her whisper, "Oh boy. Yeah."

I chucked a bit to see her so happy.

Elsie looked at me, "You're so good."

She looked radiant.

I said, "You look so pretty when you smile. You should smile more." I turned to walk towards the door. "Okay. Bye-bye."

Elsie said softly, "I missed you, too."

I asked, "Can you say goodbye to me?"
"Oh yes," Elsie said.
"Bye-bye."
I kissed her cheek once more. "See you later."
"Yes, and we're on our own," she paused to clear her throat, "on our place?"
My heart sank. She looked so vulnerable. Her eyes searched mine for an answer. Where was her anchor?
"Yes," I said. "You're in your place. This is your bed in your room. Okay?"
I left with the feeling that she wasn't so reassured this time.
(Hamilton 2003: 89–93)

As the above discussion suggests, it is possible to approach the study of language and Alzheimer's disease from a variety of perspectives. At the risk of being called a reductionist, I have decided to divide research activity to date into four categories and to give each category an iconic label: *the prism, the soliloquy, the couch,* and *the dance*. It is my hope that the use of these somewhat unusual-sounding terms will assist the reader in recalling the different foci highlighted by each approach. We turn now to a brief description of each perspective.

The prism

> A sunbeam enters a crystal prism and comes out a rainbow on the other side. Colorlessness explodes into an array of brilliant colors – yellow-orange-red-violet-blue-green.... Normal use of language and memory is like the colorless beam of light. It surrounds us. We're unaware of it. We take its existence for granted. Alzheimer's disease, as a prism, takes language and memory and spreads them out into all their varied facets.... Some of these facets remain intact, shining clear and strong. Others fade in their intensity and disappear from the spectrum, leaving an unfillable gap. As the disease progresses, these gaps widen. (Hamilton 2003: 107–9)

The prism symbolizes the focus of some researchers on the disembodied display of a variety of linguistic and communicative phenomena that characterize the breakdown that typically occurs as Alzheimer's disease progresses. I use the term "disembodied" to highlight the fact that – in this perspective – selected linguistic or discursive phenomena are analyzed apart from other linguistic or social phenomena, with no specification of a relationship to the speaker or the context of talk beyond possible subsequent straightforward correlations. Illustrations of phenomena that are the subject of such investigations include so-called "empty" words (e.g., *thing*, *place*, *make*, and *do*), circumlocutions, pronouns, and neologisms, formulaic language, and self- or other-repetition.

The soliloquy

> **"See now there's.. I've got to get this off."** Elsie had taken off her eyeglasses and was holding them up towards the light that was streaming in through the lounge window. She stared intently at the lenses and then brought them down into her lap. Steadying the frames with her left hand, she reached with her right down to the hem of her cotton dress. Bringing the fabric up to the glasses, she began a circular wiping motion. **"So I'm starting now to get it off. I don't know what they're going to that. It didn't used to hit me. There now it's going off. Yes. There it goes over down there, I'm pretty sure. Some of those, you know. And those are good. And they do pretty well, but they don't, so I'll have to get some off, I think."**
>
> She raised the glasses up to a point in front of her mouth, opened her lips ever so slightly and blew onto the left lens. **"I'll see if I'm getting of it off. 'Cause sometimes they'll go all right and other times they won't be and then let's see."** Back up towards the stream of sunlight went the glasses. A quick look through the lenses. **"Now how is this doing? It looks like it's not doing it very greasy things. One of the young men wanted to have lots of fun."** She laughs. **"So.. so on that one now I.. now I'll take a little more on this... I'll ask this one here."** Again she blew her moist breath into the left lens. **"Now."** And again. **"And then see. Then I say..look and see it now. Is that right? Is any of it off?"** (excerpt from Hamilton 2003: 70–1)

Other researchers have focused their analytical attention on the investigation of how individuals with Alzheimer's disease integrate the individual linguistic and communicative phenomena identified by the "prism" approach into a whole discourse, as seen in the stream-of-consciousness discourse captured in the excerpt just above. In this approach, linguistic phenomena are analyzed not as separate phonological, morphological, syntactic, semantic, pragmatic, and lexical phenomena, but instead are examined in light of the larger discourse in which they occur – and which they help to construct.

My use of the term "soliloquy" to characterize this approach highlights the focus on the relatively disembodied discourse produced solely by the speaker (or writer) with Alzheimer's disease, rather than on a discourse co-constructed within an interactional context. Of course – unlike a prototypical on-stage soliloquy – the discourse investigated from this perspective is indeed produced by individuals with Alzheimer's disease within an interactional context, such as a conversation. My use of the term "soliloquy" is meant simply to underscore the fact that the language of the conversational partners and/or researcher-interviewers is not included in the examination. An illustration of this type of analysis would be an investigation of how full noun phrases, "empty" words, circumlocutions, neologisms, and pronouns work together to create cohesion (or lack of cohesion) in the discourse produced by individuals with Alzheimer's disease.

The couch

Senility: The worst
is that one cannot
recognize himself.
There is a dead man
with his name

(George Oppen in Hamilton 2003: 103)

Still other researchers have focused their attention on what the language changes that typically accompany Alzheimer's disease mean to people struggling with the disease – meanings such as those expressed above by Pulitzer Prize-winning poet, George Oppen, as Alzheimer's disease threatened to cut off his life's work.[1] How is sense of self affected in response to the constant struggle to find words to express ideas and feelings? What does it mean to individuals with Alzheimer's disease when they can't seem to understand others – and others can't seem to understand them, or when they can't remember what they just said and who they said it to? How is one's identity affected by changes in ability to integrate jokes and stories seamlessly into conversation? Do these changes in language and communication ultimately lead to social isolation that, in turn, may alter an individual's sense of worth?

[1] From George Oppen's unpublished notes in the George Oppen Archive of the Mandeville Department of Special Collections, University of California, San Diego.

Here the "couch" – referring to the stereotypical couch found in a psychiatrist's office – is meant to symbolize this approach's focus on personal meaning-making, sense-making and identity construction carried out by individuals with Alzheimer's disease. This focus examines ways in which language is used to display the (potentially) changing identity of the Alzheimer's patient or, alternatively, looks at ways in which breakdowns in language over time are responded to by the patient in terms of meaning and self. Illustrations of this type of analysis include discussions of the "construction and deconstruction of self" (Sabat & Harré 1992) and Kitwood's (1997) theory of "person-centered care" where research focuses relatively more on the affirmation of personhood and less on the disease itself.

The dance

> Talking is like dancing. Dancers who are sensitive to what their partners can and cannot do can make important choices, choices that can make their partners look elegant and agile or ones that can make their partners look silly and uncoordinated. How caregivers talk with their loved ones can help or hinder them. (Hamilton 2003: 8)

The fourth group of researchers within the area of language and Alzheimer's disease focuses its attention on the interaction between language users and the ways in which interlocutors influence each other – both socially and linguistically.

My use of the term "dance" is meant to highlight the interaction as *process* rather than *product*. In this approach, researchers examine moment-to-moment turns or moves that display interlocutors' meaning-making and relationship-building as these emerge across the interaction. This is crucially different from the discourse approach characterized by the "soliloquy" above which focuses solely on the language produced by the individual with Alzheimer's disease. In the dance, the interdependence between or among partners is accentuated. This approach can focus relatively more on language or on identity construction. Effects on language can include linguistic accommodations made by healthy conversational partners (e.g., asking yes–no questions instead of open-ended questions) in conversations with individuals with Alzheimer's disease (see Hamilton 1991); effects on identity can include positioning moves that construct identities as patient or peer in conversation (see Hamilton 1996).

Where we're headed

In his discussion of the relation between linguistics and the mind, Chafe (1994) argues convincingly that no single approach can be understood to be inherently the correct one. In his opinion, all types of data "provide important insights, and all have their limitations" (1994: 12). Each methodology makes a contribution, but "none has an exclusive claim on scientific validity" (1994: 18). It should be clear from the sweeping review above that – here too – multiple disciplinary perspectives are necessary to a fuller understanding of the multifaceted nature of the relationships between language and Alzheimer's disease. As we move forward in our work in this area, then, there is no room for argument regarding whose paradigm should dominate. Individual studies will, of course, take place from particular perspectives, but no disciplinary approach should be excluded *a priori*, as this will almost certainly result in a less-than-complete ultimate understanding of the issues. The challenges are far too complex to be understood by looking through one set of filters. Future research should be centered on language, but radiate out to take into account insights from intersecting fields such as sociology, ethics, health communication, psychology, anthropology, and so on.

The fine contributions to *Alzheimer Talk, Text and Context* all focus on language – language produced by individuals with Alzheimer's disease, language produced by caregivers or other conversational partners, language of a website intended for individuals with Alzheimer's disease and their caregivers, and computer-simulated conversations between caregivers and those with Alzheimer's disease. But the specific approaches to this language differ quite significantly. While one chapter examines language used by a bilingual with Alzheimer's disease, another looks at possible ethnic influences on discourse practices in opening a conversation. Still another discusses language-related difficulties identified by family caregivers. A glance at the contributors' disciplinary and institutional affiliations gives us insight into what underlies these differences.

When taken together, these contributions respond to the appeal by Coupland *et al.* (1991:4) who argued that there is "an obligation upon socially based research to redress the balance and move away from the cognitive and psycholinguistic concerns that have come to dominate the literatures." Although Coupland *et al.* were speaking about the field of language and aging in general, the same can be said about its subfield of language and Alzheimer's disease. These contributions sketch out a robust agenda for next steps in language and Alzheimer's research and help us look optimistically and systematically toward the future. Their arguments and analyses make clear that forward movement needs to be made on two fronts: basic research and application. After little more than two decades of work in this area, there is much still to be discovered about Alzheimer's effects on language, about the effects of language problems on issues of identity and self-esteem, about interactional effects of one conversational partner's linguistic decisions on another, and on and on. On the other hand, what we do know already from basic research needs to be translated into effective training models to facilitate communication in a variety of relevant real-life contexts.

In what follows I highlight key recommendations for future research on topics at the intersection of language and Alzheimer's disease. Some recommendations relate to the continuation of research areas that are already well underway – either due to work by contributors to this volume or by other scholars in the field. Other recommendations relate to areas of research that are in their infancy – again either due to work presented here in this volume or by previous work. Still other recommendations relate to research not yet begun, but that were triggered by ideas and approaches used in the contributions to this book. My recommendations for basic research are organized under the following four headings: (1) research subjects; (2) contexts of language use; (3) theoretical frameworks; and (4) units of analysis.

Research subjects

As studies both within and outside this volume have shown, wide variation exists within language used by individuals with Alzheimer's disease. Factors such as familiarity with conversational partner, physical location, time of day, and activity can all influence an individual's use of language. The pacing of language decline can vary greatly from one individual to another. And even a single individual's language difficulties may improve and decline from day to day, although most researchers agree that the sequence of decline seems to be similar for most individuals with Alzheimer's disease.

Researchers deal with the issue of variation in different ways. Often researchers argue that the best way of compensating for wide variation is to include very large numbers of subjects. The large numbers are seen as means to greater generalization of the findings of the study; i.e., in a larger study, it is more likely that researchers will be working with a set of individuals who represent the larger population in relevant ways. In a case study or one involving very few subjects, it is more likely that the individuals will not represent the larger population in these ways. On the other hand, proponents of case studies and small-scale studies argue that the extreme variation that exists within the population makes it likely that large-scale studies simply average out these large differences, and that the averages found, therefore, are actually not representative of large numbers of the population in any meaningful way. Case studies and small-scale studies are seen as being able to investigate in more in-depth fashion the interrelationships among a variety of linguistic and social factors, leading to well-grounded research questions and methodologies that can be used in subsequent larger-scale studies.

Regardless of whether relatively more in-depth analyses are undertaken with fewer subjects or relatively more superficial analyses are undertaken with a great many more subjects, it is clear that the only differentiation within the Alzheimer's population made by most research to date has been differentiation related to the severity of the disease. Most research has looked at language changes as being related to changes in the brain and, only to a lesser degree, to changes in the interactional context. Contributions to this volume suggest that we are ready to take the next step: it is time to begin to design studies that differentiate the Alzheimer's population in (at least) the following ways.

- **native language(s)**: Does Alzheimer's disease differentially affect different native languages, in terms both of spared and deteriorating abilities? What about signed languages? Are an individual's non-native languages affected by Alzheimer's disease in different ways from that individual's native language? What about bilingualism? What about the length of time a language has been spoken or not spoken? What about the level of proficiency reached in the languages? Does a match or mismatch of regional dialect affect the quality of interactions between individuals with Alzheimer's disease and their conversational partners?
- **profession**: Does Alzheimer's disease differentially affect individuals from different professions? Are individuals who worked significantly more with language in their profession differentially affected by Alzheimer's disease?
- **education**: Does level of education affect language changes related to Alzheimer's disease?
- **sex**: Do language changes that accompany Alzheimer's disease play themselves out in the same way for females as for males? Do females and males use the same coping strategies in response to such changes? Do these language changes affect females and males in different ways

regarding identity issues? Do researchers/interviewers speak differently with male and female individuals with Alzheimer's disease?
- **ethnicity**: Does Alzheimer's disease differentially affect individuals from different ethnic groups? How does an ethnic group's beliefs regarding cognition, aging, dementia, Alzheimer's disease, and health in general affect its members with Alzheimer's disease? Does a match or mismatch of ethnicity affect the quality of interactions between individuals with Alzheimer's disease and their conversational partners?

Contexts of talk

Studies of language and Alzheimer's disease typically examine language used by and with individuals with Alzheimer's disease within one or more of the following contexts: (1) standardized testing within a clinical setting; (2) interviews or conversations with a researcher, and (3) interactions with peers (such as support groups or social activities) that are "listened in on" by the researcher. Since differences inherent in these interactional contexts can result in differences in the language produced and comprehended, some researchers have identified these different contexts as being at least partially responsible for contradictory findings across studies (see Light 1993 & Melvold *et al.* 1994).

In addition to the contextual differences related to activity and purpose noted above, important sociolinguistic work by Ramanathan (1997) has identified the relationship between conversational partners as an influential factor on language used by individuals with Alzheimer's disease. So that researchers will have sufficient opportunities to study similarities and differences in language used by individuals with Alzheimer's disease across systematically different contexts, I recommend recording interactions that go beyond the typical extended conversations or interviews to include those with a wider range of purposes, lengths, and interlocutors:

- speaking with friends and family members
- speaking with people unfamiliar to the individual with Alzheimer's disease, such as waitresses, cashiers, or new physicians
- speaking with other individuals with Alzheimer's disease, both familiar and unfamiliar
- speaking just to be sociable: over lunch with friends, meeting new people at a friend's party
- speaking to accomplish routine, unproblematic tasks, such as ordering food in a restaurant or buying stamps at the post office
- speaking to accomplish important tasks, such as speaking with a physician or working through a problem in billing at the pharmacy

- representing oneself vs. being accompanied by a spouse in formal and informal contexts
- speaking within a testing situation
- speaking within a learning context (foreign language, photography, cooking, etc.)
- speaking or writing as part of artistic or creative process (poetry, short stories)
- speaking to reminisce (life stories or narratives)
- writing or reading: for pleasure, as part of testing, or as coping or compensatory strategies

Theoretical frameworks

In order to have a solid basis upon which to see interconnections within synchronic studies (snapshots) as well as across time in longitudinal studies, and to pinpoint areas that are sensitive to contextual variables, such as task, physical contexts, and interlocutor characteristics, it makes sense to work with a robust and systematic theory of natural language use. Due to space constraints, I will mention only three possibilities, Schiffrin's (1987) model of local coherence in discourse, Halliday's (1994) systemic-functional linguistics, and The Council of Europe's (2004) Common European Framework of Reference for Languages, although there are certainly other suitable choices.

Schiffrin's interactive model identifies five components which work together to create local coherence in a discourse.

- **exchange structure**: discourse phenomena that fulfill the mechanical requirements of talk, such as turns-at-talk and adjacency pairs (Sacks *et al.* 1978 and Schegloff & Sacks 1973)
- **action structure**: actions that are carried out in talk and the situation of these actions with regard to previous and subsequent ones (Austin 1962, Searle 1969)
- **ideational structure**: propositions, or ideas, in discourse and the relations that hold between ideas, including cohesive relations (Halliday & Hasan 1976), topical relations (see Brown & Yule 1983 and Chafe 1994) and functional relations (such as relations between propositions in a narrative or an argument)
- **information state**: organization and management of knowledge and meta-knowledge (knowledge about knowledge) in the discourse
- **participant framework**: relationships between interlocutors in a conversation, as well as the relationships to what they are saying and doing in the discourse

Halliday's (1978, 1994) systemic-functional linguistics analyzes language in terms of four strata: context, semantics, lexico-grammar, and phonology-graphology. For our purposes here, Halliday's notion of context is most critical. Context is differentiated in the following way:

- **Field**: the ideational content construction
- **Tenor**: the management of interpersonal positions, roles, and faces
- **Mode**: the more formal procedural dimension of interaction

Nold (this volume) introduces the intriguing notion of using The Council of Europe's (2004) Common European Framework of Reference for Languages developed in thirteen languages by the Association of Language Teachers in Europe (ALTE). This framework uses "can-do" statements to describe what language users can typically do at six language levels referenced to four contexts (general language abilities, social and tourist typical abilities, work typical abilities, and study typical abilities). Although at first blush it may seem quite unusual to be tracking the language use of individuals with Alzheimer's disease within a framework meant for tracking progress of language learners, Nold's point about focusing on what the Alzheimer speaker *can do* as opposed to studying only what has deteriorated is a valuable one. The Council of Europe's focus on functional language abilities related to real-life contexts is also potentially useful in tracking the well-being of individuals with Alzheimer's disease within their lifeworlds.

It is easy to see how relationships between relative strengths and weaknesses in language used by individuals with Alzheimer's disease – as well as subsequent changes over time and across contexts – could be integrated into Schiffrin's, Halliday's, or The Council of Europe models. The point is to choose a model of natural language use that allows the identification and categorization of relevant language characteristics, so that relationships, patterns, and trends can be discerned. In Hamilton (1994a: 19–29) I first relied on Schiffrin's model to help organize the findings reported in my review of the literature on language and Alzheimer's disease. Halliday's theory later proved to be a useful way of tracking the differential decline in Elsie's displayed abilities to take the role of the other in conversation (Hamilton 1994a: 41–2).

Once a robust, multi-faceted, multi-component approach to discourse has been selected to serve as a backdrop, the researcher is ready to move ahead to an investigation of the specific area of his or her interest that intersects with the more general areas of language and Alzheimer's disease. The number of possibly relevant theoretical frameworks at this point in the research program is, of course, too many to do justice to in this chapter. A very select number of frameworks will need to suffice to give the reader a sense of the possibilities.

- **person-centered care** (Kitwood 1997): Language is analyzed from this perspective in at least three distinct ways: (1) in an effort to accentuate what is positive rather than negative, spared language abilities can be identified; (2) in an effort to turn away from a perception of "loss of self" in Alzheimer's disease, language used by the individual with Alzheimer's disease to construct identities other than as patient can be examined; and (3) in focusing on the caregiver as facilitator of improved and expanded interactions with the individual with Alzheimer's disease, the language the caregiver uses is of obvious analytical interest.
- **communication accommodation theory** (Coupland *etal.* 1988): From this perspective, language of conversational partners is examined for evidence of convergence or divergence; i.e., to see whether they are becoming relatively more or less alike discursively as the interaction continues. This perspective can be a useful tool for investigating and identifying communicative strategies used by caregivers of individuals with Alzheimer's disease in pursuit of "caregiver as facilitator" within Kitwood's model of person-centered care (see above).
- **interactional sociolinguistics** (Gumperz 2001; Schiffrin 1994): Language is analyzed from this perspective to understand how individuals with Alzheimer's disease and their interlocutors make sense within emergent interaction. Attention is paid to the role of background knowledge and knowledge about contextual frames and speech activities in this sense-making. Fine-grained analysis of linguistic and paralinguistic features uncovers how conversational partners display their communicative intentions, draw communicative inferences, signal social relations, and co-construct the ongoing activity.
- **identity construction** (Gergen 1991, 1999; Goffman 1959) From a social constructivist perspective, language use both reflects and creates an individual's identity. "One's identity is continuously emergent, re-formed, and redirected as one moves through the sea of ever-changing relationships" (Gergen 1991: 139). Language can be examined for evidence of types of identities being constructed by the individual with Alzheimer's disease for him- or herself, as well as by healthy interlocutors for the individual with Alzheimer's disease.
- **positioning theory** (Davies & Harré 1990; Sabat & Harré 1992; Bamberg 1997): From this perspective, conversational language can

be examined for evidence of self- and other-positioning as it emerges across the interaction. The connection of these positionings to multiple personal storylines allows the analyst to disentangle multiple identities (e.g., parent, child, patient, professional) as they are enacted in fleeting ways over time. Bamberg's work enables fine-grained investigations into positioning within personal experience narratives, where one can look at three levels: (1) positioning among storyworld figures; (2) positioning of narrator to conversational partners; and (3) positioning of narrator to self ("Who am I?").

- **politeness theory** (Brown & Levinson 1978, 1987): From this perspective, language is understood as evidence of the balancing act necessary to sustain both positive and negative face in interaction (the need to be liked vs. the need to be autonomous). In the execution of acts that would threaten either positive or negative face of self or other, speakers assess the social distance and the relative power differential between listeners and themselves, along with the level of imposition related to the face-threatening act, and select appropriate language to carry out the act. Language analysts then can use this selected language as a window to speakers' perceptions of relative social distance, power, and imposition. Politeness theory can be successfully used to examine identity construction and positioning, as well as to track shifts in communicative competence and social acuity of speakers traveling the path of Alzheimer's disease.

Units of analysis

Depending on what theoretical framework guides the study, a variety of units of analysis can be chosen. Aspects of language that have received a good deal of analytical attention in Alzheimer's disease include: "empty" words, circumlocutions, neologisms, repetition/perseveration, formulaic language, conversational routines, social talk, topic maintenance, questions, and repair. It would be an exercise in futility to try to list all potentially relevant units in studies of language and Alzheimer's disease, as these are the building blocks for any linguistic or discourse analysis. A more useful approach at this point may be to recall the four general categories outlined above – the prism, the soliloquy, the couch, and the dance – and to think of the relevance of analytical units to solving research problems from those perspectives.

Some units may be called on to help researchers follow Obler's (1983: 271) call for study of "the process of deterioration in order to discover the semiotic hierarchies of pragmatics" by pinpointing the sequencing of language and communicative breakdown in Alzheimer's disease. Each illustrative item in the following list can be examined to track language differences related to the trajectory of the disease, activity types at the same phase of the disease, differences in conversational partners, and language typology. Other units may be useful in uncovering the specific composition of the most difficult communicative problems identified by caregivers in their interactions with Alzheimer's speakers (see Byrne and Orange, this volume). Still others may work together to characterize language use related to the construction and maintenance of particular positions or identities in interaction, such as the problematic shifts between "parent" and "child" as mentioned by Brewer (this volume). With these in mind, we can more readily recognize that one and the same unit may be selected in the service of studies emanating from different perspectives.

- **phonology/orthography**: sound systems; spelling; handshape, location, movement (for signed languages)
- **morphology**: inflection; derivation; compounding
- **lexicon**: word classes; lexical richness; "empty" words; discourse markers; professional jargon; deictic terms
- **syntax**: grammatical accuracy; complexity; clause types; length of clauses
- **semantics**: sense; reference; denotation; connotation; semantic properties and relationships
- **pragmatics**: implicature; presupposition; speech acts; cooperative principle
- **discourse**: coherence; cohesion; topic initiation, development and closure; speech acts; turn-taking; backchanneling; face management; rapport building; conversational style; intersubjectivity
- **genre**: conversation; narrative; life story; letter; essay; autobiography; poem
- **nonverbal communication**: paralinguistics; gestures; facial expression; gaze; posture

In closing

"We must learn to live in a new way"
(George Oppen in Hamilton 2003: 108)

For individuals with Alzheimer's disease and for those who care for them – both personally and professionally – change is inevitable. Changes in ability to find the desired words in conversation. Changes in being able to recall who the visitor is. The contributions to *Alzheimer Talk, Text and Context* are full of such illustrations. Byrne and Orange (this volume) have identified those changes that are most problematic for caregiver–patient relationships – repetitive questions, word-finding difficulties, "nothing come back" in conversation, the same story or piece of information being told over and over again, and difficulty initiating and following a conversation. Equally important, however, is the encouraging thread underlying so many of these contributions: that we must never lose sight of the human being at the center of this disease. We must continue to try to find connections between these communication changes and emotional, physical and spiritual ones.

In response to the everyday challenges of Alzheimer's disease, poet George Oppen wrote the words, "We must learn to live in a new way" to his wife and caregiver in their dialogue journal. This personal mandate within the Oppen family can serve as a well-worded challenge to those of us working at the intersection of language and Alzheimer's disease. Following Kitwood (1997), maintaining a person-centered focus compels us to continue basic research towards the understanding of these changes. It also – crucially – challenges us to pause long enough along the way to translate our findings into applications that help caregivers and individuals with Alzheimer's disease find and feel comfortable with their "new normal" status – by reducing frustration levels, lifting their spirits, and encouraging what Ryan *etal.* (this volume) have called the "warmth of reciprocity." This volume's contributions suggest to us that these important activities can and should be carried out on multiple fronts: in face-to-face support groups and workshops, in the production of videos and CDs, in the development of computational modeling, and in improved Web-based programs.

The painter's canvas that had been nearly blank three decades ago now evidences vivid strokes in several areas. The background connecting all corners of the canvas has been lightly sketched out. Clusters of carefully detailed work can be found. Studies connecting some of these clusters are underway. The only way to get closer to completing the painting, however, is through continued collaborative research from multiple perspectives. To this end, I applaud both Davis and Moore for designing and executing their large-scale data collection effort and the contributors to this volume for taking on the challenge of working from a variety of perspectives with this common set of interactions. Forging ahead in this – and similar – cross-disciplinary community efforts will allow us to continue solving the myriad mysteries that lie at the intersection of language, Alzheimer's disease, and society.

References

Austin, J. (1962) *How to Do Things with Words*. Cambridge, MA: Harvard University Press.

Bamberg, M. (1997) "Positioning between structure and performance". *Journal of Narrative and Life History* 7: 335–42.

Bayles, K. (1979) "Communication Profiles in a Geriatric Population." Unpublished Ph.D. dissertation. Tucson, AZ: University of Arizona.

Bayles, K. (1982) "Language function in senile dementia." *Brain and Language* 16: 265–80.

Bayles, K. (1984) "Language and Dementia". In A. Holland (ed.), *Language Disorders in Adults*, San Diego, CA: College-Hill Press, 209–44.

Boden, D. & Bielby, D. (1986) "The way it was: topical organization in elderly conversation." *Language and Communication* 6: 73–89.

Brown, P. & Levinson, S. (1978) "Universals in Language Usage: Politeness Phenomena." In E. Goody (ed.), *Questions and Politeness*, 56–289. Cambridge: Cambridge University Press.

Brown, P. & Levinson, S. (1987) *Politeness: Some Universals in Language Usage*. Cambridge: Cambridge University Press.

Brown, G. & Yule, G. (1983) *Discourse Analysis*. Cambridge: Cambridge University Press.

Chafe, W. (1994) *Discourse, Consciousness and Time*. Chicago: University of Chicago Press.

Coupland, N., Coupland, J., Giles, H., & Henwood, K. (1988) "Accommodating the elderly: invoking and extending a theory." *Language in Society* 17: 1–41.

Coupland, N., Coupland, J., & Giles, H. (1991) *Language, Society and the Elderly*. Oxford: Blackwell.

Crystal, D. (1984) *Linguistic Encounters with Language Handicap*. Oxford: Blackwell.

Davies, B. & Harré, R. (1990) "Positioning: The discursive production of selves." *Journal for the Theory of Social Behaviour* 20: 43–63.

Gergen, K. (1991) *The Saturated Self: Dilemmas of Identity in Contemporary Life*. NY: Basic Books.

Gergen, K. (1999) *An Invitation to Social Construction*. Thousand Oaks, CA: Sage.

Goffman, E. (1959) *The Presentation of Self in Everyday Life*. NY: Doubleday.

Goffman, E. (1967) "On Face-Work." In Goffman, *Interaction Ritual*. Garden City: Doubleday, 5–45.

Gubrium, J. (1975) *Living and Dying at Murray Manor*. NY: St. Martin's.

Gumperz, J. (2001) "Interactional Sociolinguistics: A Personal Perspective." In D. Schiffrin, D. Tannen, & H.E. Hamilton (eds.), *The Handbook of Discourse Analysis*. Oxford: Blackwell, 215–28.

Halliday, M. (1978) *Language as Social Semiotic*. London: Edward Arnold.

Halliday, M. (1994) *Introduction to Functional Grammar*, second edition. London: Edward Arnold.

Halliday, M. & Hasan, R. (1976) *Cohesion in English*. London: Longman.

Hamilton, H. (1991) "Accommodation and Mental Disability." In H. Giles, J. Coupland, & N. Coupland (eds.), *Contexts of Accommodation*, 157–86. Cambridge: Cambridge University Press.

Hamilton, H. (1994a) *Conversations with an Alzheimer's Patient: An Interactional Sociolinguistic Study*. Cambridge: Cambridge University Press.

Hamilton, H. (1994b) "Requests for Clarification as Evidence of Pragmatic Comprehension Difficulty." In R. Bloom, L. Obler, S. DeSanti, & J. Ehlich (eds.), *Discourse Analysis and Applications: Studies in Adult Clinical Populations,* 185–99. Hillsdale, NJ: Lawrence Erlbaum.

Hamilton, H.E. (1996) "Intratextuality, intertextuality and the construction of identity as patient in Alzheimer's Disease." *Text* 16/1: 61–90.

Hamilton, H. (2000) "Dealing with Declining Health in Old Age: Identity Construction in the Oppen Family Letter Exchange." In J. Peyton, P. Griffin, W. Wolfram, & R. Fasold (eds.), *Language in Action: New Studies of Language in Society.* Cresskill, NJ: Hampton Press, 599–610.

Hamilton, H. (2003) *Glimmers: A Journey into Alzheimer's Disease.* Ashland, OR: RiverWood Books.

Irigaray, L. (1973) *Le langage des déments.* Le Hague: Mouton.

Kempler, D. (1984) "Syntactic and Symbolic Abilities in Alzheimer's Disease." Unpublished Ph.D. dissertation. Los Angeles: UCLA.

Kitwood, T. (1988) "The technical, the personal, and the framing of dementia." *Social Behaviour* 3: 161–79.

Kitwood, T. (1997) *Dementia Reconsidered: The Person Comes First.* Philadelphia: Open University Press.

Light, L. (1993) "Language Change in Old Age." In G. Blanken, J. Dittmann, H. Grimm, J.C. Marshall, & C.W. Wallesch (eds.), *Linguistic Disorders and Pathologies: An International Handbook.* Berlin: Walter de Gruyter, 900–18.

Lubinski, R.B. (1976) "Perceptions of Oral–Verbal Communication by Residents and Staff of an Institution for the Chronically Ill and Aged." Unpublished Ph.D. dissertation. Columbia University Teachers College.

Lubinski, R.B. (1981) "Language and aging: An environmental approach to intervention." *Topics in Language Disorders* 1: 89–97.

McTear, M. & King, F. (1991) "Miscommunication in Clinical Contexts: The Speech Therapy Interview." In N. Coupland, H. Giles, & J. Wiemann (eds.), *"Miscommunication" and Problematic Talk.* Newbury Park: Sage, 195–214.

Melvold, J.L., Au, R., Obler, L.K., & Albert, M.L. (1994) "Language during Aging and Dementia." In M.L. Albert & J. Knoefel (eds.), *Clinical Neurology of Aging* (second edition). NY: Oxford University Press, 329–46.

Obler, L. (1983) "Language and Brain Dysfunction in Dementia." In S. Segalowitz (ed.), *Language Functions and Brain Organization.* NY: Academic Press, 267–82.

Obler, L.K. & Albert, M.L. (eds) (1980) *Language and Communication in the Elderly: Clinical, Therapeutic, and Experimental Aspects.* Lexington, MA: D.C. Heath.

Ramanathan, V. (1997) *Alzheimer Discourse: Some Sociolinguistic Dimensions.* Mahwah, NJ: Erlbaum.

Ripich, D. & Terrell, B. (1988) "Patterns of discourse cohesion and coherence in Alzheimer's Disease." *Journal of Speech and Hearing Disorders* 53:8–15.

Sabat, S. & Harré, R. (1992) "The construction and deconstruction of self in Alzheimer disease." *Ageing and Society* 12: 443–61.

Sacks, H., Schegloff, E., & Jefferson, G. (1978) "A Simplest Systematics for the Organization of Turn-Taking for Conversation." In J. Schenkein (ed.), *Studies in the Organization of Conversational Interaction,* 7–55. NY: Academic Press.

Schegloff, E. & Sacks, H. (1973) "Opening up closings". *Semiotica* 8: 289–327.

Schiffrin, D. (1987) *Discourse Markers.* Cambridge: Cambridge University Press.

Schiffrin, D. (1994) *Approaches to Discourse.* Oxford: Blackwell.

Searle, J. (1969) *Speech Acts*. Cambridge: Cambridge University Press.
The Council of Europe (2004) *The Common European Framework of Reference for Languages*. Cambridge: Cambridge University Press.
Ulatowska, H.K. (ed.) (1985) *The Aging Brain: Communication in the Elderly*. San Diego CA: College Hill Press.

Index